13.56
CIVIC GARDEN CENTRE LIBRARY
3002937

D1058046

GLADIOLI FOR EVERYONE

Gladioli for Everyone

John Bentley Garrity

DAVID & CHARLES
NEWTON ABBOT LONDON VANCOUVER

0 7153 6803 6

Set in 11pt on 13pt Linotype Plantin and printed in Great Britain by Latimer Trend & Company Ltd Plymouth for David & Charles (Holdings) Limited South Devon House Newton Abbot Devon

Published in Canada by Douglas David & Charles Limited 132 Philip Avenue North Vancouver BC

Contents

List of Illustrations

Preface

I am not a trained horticulturalist, geneticist, botanist or any other kind of scientist; nor am I engaged in wholesale or retail flower-trading. However, I love all sort of flowers, especially the gladiolus in its various forms. It is part of the intention behind writing this book to demonstrate that the ordinary gardener can, if he has the will to do it, learn enough to become reasonably expert about any genus—certainly to grow it well, exhibit it in competition at national standards, and breed new varieties.

This book then, though specialised in the sense that it deals with one genus only, is intended for the ordinary gardener. But, with the keen show competitor also in mind, in places it offers a counsel of perfection rather than what it is absolutely essential to do. Any technical terms are explained and a handy glossary is included as an appendix. Without pretending to be a completely definitive book on the genus *Gladiolus,* it does attempt to be thorough, so that the reader should know not only what to do, but why he is doing it.

The second intention is that it should act as a reference book for all who grow, or wish to grow, gladioli. This is achieved by the grouping of material under specific headings and by a full index at the back.

The third intention I make no apology for: it is to encourage others to share the pleasure that I and my fellow-enthusiasts derive from growing, showing and breeding gladioli. There is no substitute for the actual flower; and prohibitive printing costs make it impossible to illustrate in colour the vast range of fine modern gladioli. My wish is that readers of this book will derive as much joy from the gladiolus as I have done.

John Garrity

Chapter 1

The Nature of the Plant

The cycle of the life of a gladiolus begins with the planting of what is termed a 'bulb' in the USA and Canada, but a 'corm' throughout most of the rest of the gladiolus-growing world. However handy it may be to refer to all corms as 'bulbs', there is a difference in name for good reasons and it needs to be understood why botanists insist upon it; they are not just being pedantic.

The basic difference between the two is that a corm is a swollen stem-base, whereas a bulb is composed of swollen leaf-bases. Both have the same function of food storage, but there are other differences both in structure and in the way the developing plant grows with the help of these food reserves.

Every year the planted corm dies by withering away as it helps to feed the current new growth, and a new (usually larger) corm is formed above by lateral (sideways) swelling of the plant's stem. In the case of a bulb, the use of the food reserves also causes withering of the previously fat leaf-bases, which eventually become the papery scales seen on its outside. However, the bulb does not become smaller, because new food reserves are laid down inside it at the base of, and by the action of, the current season's leaves. Right at the heart of a bulb, which is a much modified bud, is an embryonic (undeveloped) shoot that includes a complete flower for the next year. With the onset once more of favourable growing conditions, the embryo is fed by the bulb, grows into a flowering plant, and so continues yearly.

In contrast, a corm is one large food-store and virtually entirely stem-tissue, wrapped around by parchment-like scales that are the dead remnants of the previous year's leaf-bases. However, corms vary from bulbs in another way, because the floral primordia (embryonic flowerheads) will not develop until a certain amount of foliage growth has taken place. The floral primordia are not present when the corm is planted and even the leaves are represented by only the minutest structures, whereas in bulbs they are well developed.

The miracle is that inside that apparently uniform corm-tissue is a genetic code that not only directs certain cells to become roots and others to become leaves, but also determines what type of flower will eventually bloom and in what colours.

The nodes of a plant are the points on the stem from which leaves sprout. It is noticeable in the garden gladiolus how the leaves are all set low on the stem, indeed below ground-level. This is because the lateral swelling of the stem causes the internodes (portions of stem between pairs of leaves) not to elongate as they would in other plants, and so the leaves they carry always remain clustered at the swollen stem-base. Indeed, corms have been described as 'compressed shoots'; more accurately, they are shoots that do not elongate, except for their flower-spikes. Clean the dead leaf-husks from a corm and you will be able to see rings or ovals on it where the leaves were once attached.

The leaves grow up in a fan across one plane only, except for the two to five outside ones, which are not true foliage, but merely sheaths that come a little way above ground and are wrapped around the base of the foliage-leaves. The corm will show tiny buds, also along one axis. There is usually one between each pair of rings and they alternate: if the apical (most central) is to the left of the old stem, the second will be to the right, the third to the left, and so on.

This is where the corm is again different from a bulb. The number of spikes that a large corm will produce can be fairly accurately determined at planting-time. The apical bud will usually be the first to break dormancy (winter hibernation) and will always give the strongest spike. Therefore, if you want one good upright spike, especially for show purposes, care is taken not to damage the apical bud, but all the others are rubbed out with the thumb, so that you are effectively 'de-eyeing' the corm and restricting it to one shoot, into which all its energies will go.

If the second-to-centre bud is left, that will put up a second spike ninety-nine times out of a hundred.

Thereafter, the number of spikes coming from an untampered-with corm will vary with the cultivar (cultivated variety, as opposed to natural variety of a wild species). In most cultivars, if the apical and second buds start into growth, a chemical inhibitor is released that makes the others remain dormant. However, for garden and cut-flower purposes some cultivars that have a weak inhibitor mechanism and will give four or five or more stems from one large corm are being interbred to produce strains that can be relied upon to give several flower-stems from each corm. To multiply quickly stock of those with strong inhibiting tendencies, rub out the apical bud and (if there are several others) the second one too.

If not interfered with, the summer-flowering garden gladiolus will naturally break from dormancy at a specific time (March to April in the northern hemisphere), though this varies from cultivar to cultivar. Stored corms are apt to begin sprouting before one gets around to planting them. There is no great harm in this, if the sprouts do not become too long and are rising vertically from horizontally stored corms. However, at the same time the tiny root primordia around the basal plate (the central disc on the underside of the corm) will be starting into growth and could wither if left too long out of the ground and therefore not absorbing moisture.

Once the corms are in the ground and growth has begun, it is important that the roots develop quickly and penetrate a largish volume of soil. Of two distinct sets of roots, the first emerge from the planted corm's base and are the feeding-roots. They branch and re-branch, the ends being very fine and having even finer 'root-hairs' growing from specialised epidermal (surface) cells. These finer roots periodically die and are replaced by new and more extensive ones throughout the growing season. The plant draws moisture from the soil by means of these roots, and therefore takes up whatever is in solution in the soil-water, be it beneficial or injurious. Healthy root-growth requires not only the right balance of chemical intake and adequate photosynthesis (see below), but the roots must neither dry out nor reach into stagnant water and also adequate aeration is necessary to provide oxygen. Thus the total extension of the roots is not only limited by the proximity of other plants, it is also determined by the depth of well-aerated soil and the closeness to the surface that the roots can come yet still find 'free' soil-water.

The tip of the bud, as it breaks through its protective scales, has seven or eight rudimentary leaves already starting to grow in their alternate overlapping positions. The lowest internodes grow most rapidly at first, pushing the ensheathing leaves through the ground. These are often of a redder colour than the true foliage. The foliage leaves elongate in turn and emerge from the centre, curving alternately to left and right. Until they emerge, the plant is relying upon the food-store within the corm, so different sizes of corm need different depths of planting. The minerals absorbed with the soil-water can serve only a limited function until the green leaves, which are green because they contain chlorophyll, begin their own chemistry with the aid of sunlight. This is known as photosynthesis (uniting by light) and causes carbon dioxide and water to form sugar, which provides the energy for growth throughout the plant. Actually two types of chlorophyll (a and b) and carotenes (orange pigment) and xanthophylls (yellow pigment) have to be present for the chemistry to work effectively.

The veins are prominent and give a ribbed effect. In a healthy plant the leaf area between veins should be a deeper green than the veins. These stiffen the leaves with thick-walled fibrous tissue, but also act as service ducts. The xylem is the tube that brings up the water and minerals for photosynthesis; the phloem distributes the organic results (and, incidentally, any virus infection). Stomata (tiny breathing pores) on the leaf-surface enable the leaf to take in carbon dioxide and emit oxygen. Before we extend our 'concrete jungles' any farther, perhaps we would do well to consider the part that nature plays in ensuring a balance of life-maintaining oxygen in the atmosphere.

The siting of gladiolus beds where the plants will get maximum sunlight throughout their growth cycle is obviously important, because sunlight is essential to photosynthesis. However, a high rate of photosynthesis and a high rate of oxygen-release involve a great deal of transpiration (giving off of water-vapour) through these same stomata, so the better the plants are growing, the greater their water demands.

The precise time when the floral primordia begin to form at the growing tip varies according to conditions, but is quite early in any plant's life. On average one could say that the process had started about a month after the corm began to show its first true foliage leaf. This is the stage at which the number of flowerbuds is determined. One determining factor is the weight of the corm planted; another is the number of shoots that begin to grow, and the third is the cultural conditions experienced by the corm during that month. It hardly need be emphasised that good culture must begin early, first with the soil preparation before the corms are even planted.

Even as the floral primordia are differentiating themselves, the stem base is already swelling to make the new corm that will eventually be harvested. At the same time a second set of roots is developing from between the top of the old corm and the base of this new corm. These are the contractile roots and are somewhat different from the original (filiform) roots. Their surface cells are highly specialised for a specific job.

These roots are initially smooth and relatively thick, up to 7mm in diameter. They are needed to pull the new corm down to its correct depth in the soil and this, amazingly enough, is what they do. Against all the resistance of the soil they gradually draw the whole plant lower. The extended cells of the surface contract and these roots wrinkle; some of the food stored in them apparently goes to feed the new corm, for they also narrow somewhat as this corm finds its optimum level. Thus nature ensures that corms left in the ground from year to year do not eventually surface.

Whilst the true foliage-leaves are attaining their maturity, the stem with the floral primordia at the tip grows very slowly. According to the ancestry of the cultivar, a time will come for each when there is a dramatic surge upwards. Protected by enveloping stem (cauline) leaves and bracts, which are modified leaves, the young flowerhead begins to force its way up through the centre of the foliage-leaves. A thickening and rounding of the foliage-base can be felt as it does so. Usually there are seven or eight true leaves growing from a full-size corm when this happens; but, as there is a transition series to the bracts surrounding the flowers, it may appear that there are ten or occasionally even more.

The rapid ascent of the spike can be detected by daily feeling the leaves and checking how far up they have been thickened by the growing tip. It is obviously vital that the plant should not lack water at this stage of fast development. As the spike lengthens, so the individual flowers develop their own identities. The primordia are separated into a bract and bracteole (the pair of green structures that surrounds each flower) and the major components of the embryonic flower. The bracts, which are on alternate sides, overlap in the emerging spike as it pierces through the leaves and appears in the characteristic form that gave *Gladiolus* its name (see Chapter 7).

The flower-parts are meanwhile set down in the following sequence: first the stamens (the male reproductive organs), normally three but occasionally more numerous or aborting into petals; then the perianth (all the petals); and finally the carpel set (female reproductive organs). Botanists have not been able to determine whether the petals are true petals or coloured sepals; nevertheless, they are normally termed 'petals' though in technical publications and horticultural articles they may be referred to as 'perianth segments', 'perianth lobes', or even 'tepals'.

When the whole flowerhead is clear of the foliage, although the lower stem continues to lengthen, there is an accelerated growth of the flowerhead section, enabling the buds to separate in sequence, lowest first. The tip of the flowerhead, although growing as precisely as circumstances will allow it away from the centre of the Earth, will usually bend slightly in one direction and the bottom buds will also lower themselves in this direction.

The placing of a stake or cane can therefore be made as soon as either of these indications appears, without fear of damaging opening flowers.

The parts of the flower are illustrated and labelled in Fig 1, below. The flower consists of an outer whorl of three petals (sometimes termed sepals) and an inner whorl of three petals which are not fused together (as in the narcissus), but technically constitute the corolla (little crown). These petals are nature's

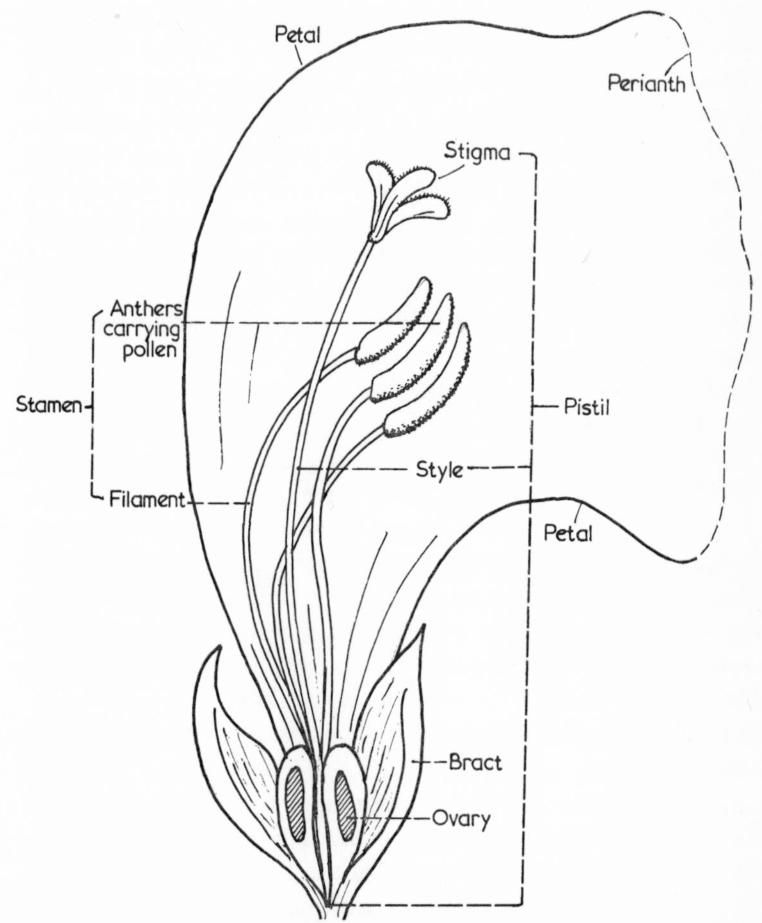

Fig 1 Parts of the flower in cross-section

device for signalling to pollinating insects that a plant has nectar for them (and is, incidentally, ready for fertilisation). It is man's good fortune that they happen to have shapes and colours attractive to him also. The insect does not come to collect pollen; it hopes to find nectar in the throat of the flower. Nature, too, works in mysterious (and highly effective) ways, its wonders to perform. From the throat there arise three slender filaments, each carrying a two-lobed anther. By about its second day open, the flower has a centimetre of ripe pollen grains along each lobe and these rub off or drop on to intruding insects, to be carried off to another flower on their journey. If this happens to be a gladiolus-flower, then the chances are that fertilisation will be effected. Above and behind the three stamens rises a style terminating in a three-lobed stigma. These lobes are also centrally grooved. When the stigma is ready to receive pollen, the minute 'hairs' (papillae) on the stigma become sticky, causing pollen carried by visiting insects to adhere. To make doubly sure of some pollination, about three days after the flower has opened the stigma arms flatten out and also separate widely from each other, thus presenting a large surface area. At the same time, the style curves forwards and downwards into its own anthers to enable self-pollination to take place.

Once ripe pollen adheres to the stigma, a chemical reaction 'triggers off' each grain and causes it to grow a very fine tube down through the style to the ovary section of the pistil, where each ovum (egg) is fertilised by a different pollen grain and will eventually become a separate seed. When all the ova are so fertilised, the purpose of the flower has been attained. Whether or not all the ova in one ovary do become fertilised, the flower will wilt and collapse after a few days, whilst others above are flaunting their blandishments to ensure that the genus will continue.

However, when those seeds are ripe, they will—unless the fertilisation has been restricted within one unvarying species—be the embryos of a whole group of somewhat different gladioli. The original plant has yet a further way of ensuring that its own individual self is perpetuated, in case the new corm(s) it forms should rot or otherwise be destroyed. Short stolons grow from the base of the new corm and clusters of cormlets develop on the ends. Once these are detached from the mother corm, they will send up fine grass-like leaves in the spring and eventually grow into flowering-size corms. A few may do this whilst still attached to the parent plant, but most are usually inhibited from doing so, as nature's way of preventing overcrowding.

All the vegetative reproductive material from any corm—its new corms and cormlets—constitute a clone, that is, they all have the same genetic make-up and will grow into flowers identical to that of the parent corm, except for an occasional 'sport' (a chance mutation that results in a variation, usually of colour). This is the way that every new gladiolus introduction has to be propagated, by growing on all corms and at least the largest of the cormlets.

The colour of each cultivar (cultivated variety) is determined largely by hereditary factors, involving not just the colours of its immediate parents, but colours and colour combinations made possible through genes passed down along its whole ancestry (see Chapter 8 for more detail). However, the chance element acts more frequently upon colours, which seem to be the least stable aspect of the genus. Gladioli range through numerous shades from white to saturated reds and purples that look virtually black. Almost all other colours are represented, including tan and mahogany shades; blue and green are the only two major colours that cannot as yet be found in pure form. This is because of the chemistry of the genus. The pigmentation (colouring) is determined by the presence or absence of anthocyanins, which give tones ranging from violet through lavender, lilac, orchid, pink, and rose, to red; and anthoxanthins, which give every tint and shade of yellow, from ivory and cream to the deepest golden yellow. Some green appearance occurs in basically white flowers ('Green Ice', 'Green Willow' etc) and also in basically yellow flowers ('Green Woodpecker' etc) and this has been recently increased by breeding, though it is one of the quickest colours to fade in sunlight, which makes one suspect that chlorophyll in the cell-sap or elsewhere is responsible for the initial greenness, seen in the yellow blotches of many cultivars as the flowers first open.

The anthocyanin delphinidin, present in true-blue flowers, such as blue delphiniums, is missing from *Gladiolus* chemistry. The nearest chemical present is the malvidin in the violet and purple cultivars. This is nearly related chemically to delphinidin, but unless a natural or artificially induced mutation of it occurs, there will probably never be a perfectly blue gladiolus.

As the flowers die, so the ovary enlarges and allows the perianth, anthers, and style with stigma to wither and drop or blow off. The three fronds of the stigma lead to three chambers in the ovary, each subdivided into two, so that six rows of seed are swelling within the seedpod. In time this grows to look like a miniature rugby football about 40 to 50mm long. Eventually it splits down its 'seams' and the three chambers part, opening outwards and downwards to allow the wind to disperse the seed. Each seed is a small globe inside a roughly circular papery brown wing, one of nature's devices for ensuring that seed-dispersal is over a wide area.

If one of those seeds finds suitable conditions for its growth, the wing-tissue rots away. The specialised cells that form a jacket around the seed prevent it from rotting and in early spring it germinates, putting down a small root radicle and putting up a tiny grass-like leaf. Within three seasons this new plant will be in flower—so we have come full cycle.

Chapter 2

Culture

Fortunately, a suitable site for growing gladioli can be found in almost every garden, on any allotment, and in every park. Like any other plant, the gladiolus will flourish best in conditions ideally suited to the individual cultivar; but the differences in performance among the various cultivars is minimal on any given site and the plants will grow quite well under a varying range of conditions. However, to reap the best harvest one has to approach as near as possible to the optimum conditions for the modern outdoor summer-flowering range of gladioli.

These flowers, with their largely South and East African ancestry, naturally do best on sites receiving full sun throughout the growing season, not just during the months that the plants are in bloom. Areas that are partially shaded during March and April, late August and September are the second-best choice; but on no account should corms be planted under trees or where houses, garages, sheds, tall walls etc throw solid shadows for a large part of the day during April to September. From the time the leaves begin emerging, they need not merely daylight but sunlight for the chlorophyll in them to be best stimulated into performing its life-giving function of transforming chemicals in solution imbibed by the roots into nutrients within the plant creating growth.

For cut-flowers and exhibition purposes, the corms should be planted in adequately spaced rows (see p 39 for distances for various types); for garden decoration, groups of corms may be planted in the herbaceous border, in special small beds with under-plantings of shallow-rooted low-growing summer flowers, on the south side of the house but at least 1ft from the wall, or in any warm corner of the garden that is not too lengthily overshadowed. Personal choice will decide the arrangement in these cases, but a small circle of from five to nine corms (according to cultivar size) with one corm in the centre is a familiar and satisfying pattern. Such gladioli should be of one colour, though not necessarily of the one cultivar, and each neighbouring circle should preferably be of strongly contrasting colour.

One idea you might try experimenting with—if you are planting in front of a building or fence—is to place the corms in zigzag formation, each 'tack' being three, four, or five corms long (see Fig 2), according to the space available. Where more than one row can be accommodated, the effect will be enhanced. This gives much scope for blending and contrasting colours, and is also very suitable for that mixed batch, the colours of which you cannot be sure about. In a border facing the road or a lawn, a zigzag of the shorter smaller-flowering gladioli, either primulinus or non-primulinus, may be grown in front of the taller large-flowered ones; but the front row will then bloom earlier than the back row, unless a very careful choice of cultivars is made (see Chapter 5 for suggestions).

For corners and small casual clumps, the smaller-flowered types are best, as they rarely need staking in sheltered spots and will grow to their right proportions in soil that has not been heavily manured or otherwise well fed. Do not, however, expect any gladioli to compete with hedge roots. Hedges are gross feeders; where they have been trimmed, their roots usually reach farther into the garden than the measurement of the hedge's height. If you unearth any hedge roots when planting gladioli, you are already too close to give the gladioli a fair chance to show you how well they can perform.

SOIL PREFERENCES

In brief, the modern gladiolus will grow best in medium loam containing plenty of humus, with a good depth of friable organic soil, no hard pan where water can collect to exclude air from the roots, and a pH value of 6·5 to 6·8, ie slightly acid. Any gardener or allotment-holder enjoying such conditions already is no doubt counting his blessings; but they can gradually be created in specific areas by determination and patient application, even when the initial 'garden' is far from the desired ideal.

The first thing to ensure is adequate drainage; though the gladiolus requires a copious supply of water during its growing season, it means certain death to the plant if the fibrous tips of

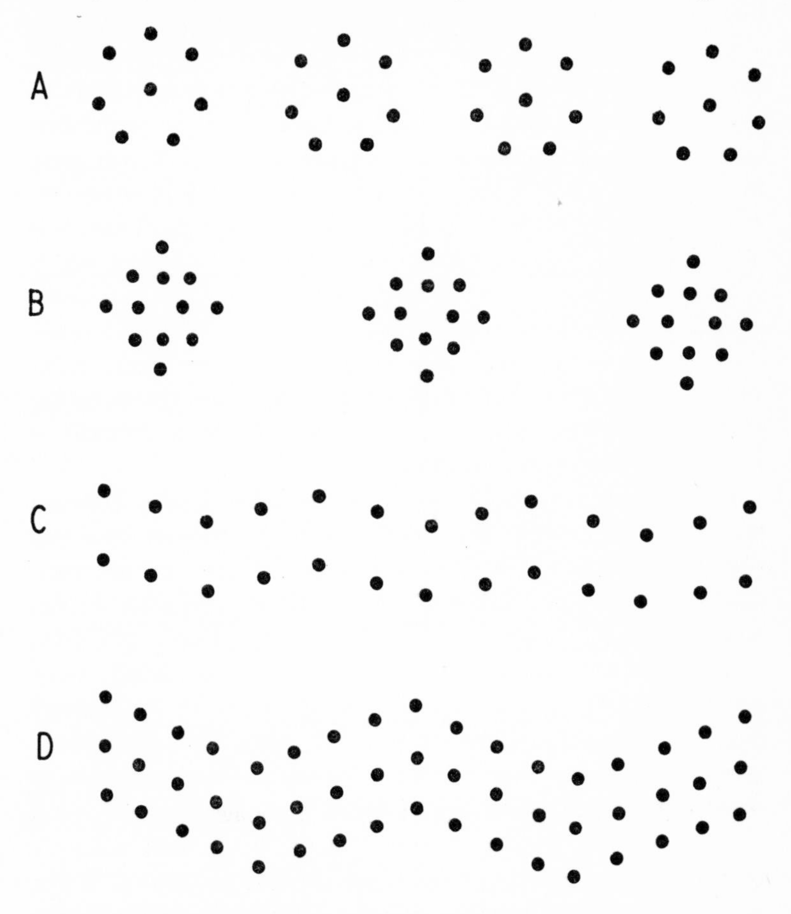

Fig 2 Planting patterns for beds or borders

its roots are immersed for long in air-excluding water. Gladiolus roots start growing nearly 6in below soil level and would like to grow about 18in. If you have any doubts about drainage conditions down to 2ft deep in your soil, an exploratory double-width trench can be dug across the area intended for gladioli. Should a hard pan of compressed clay show up before a 2ft depth is reached then the whole plot will need to be double-dug—by the method illustrated in Figs 3, 4—and the next 6in (150mm)

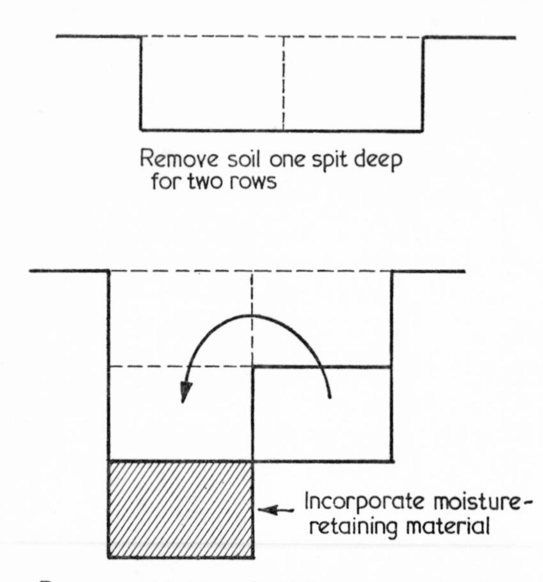

Figs 3/4 Double-digging

of the third spit broken up to have drainage material incorporated with it. This would be whatever is cheapest and most easily available to you: weathered clinker, broken-up bricks, pebbles or gravel, smashed tiles or slates, coarse sand, vermiculite, broken china—any combination of materials that will help the water to flow without introducing anything chemically injurious to plant-life. If essential, a ditch would need to be dug at the lowest end of the plot, possibly with a soak-away in an area of very heavy clay.

The very few readers who will find the above procedure necessary should take the opportunity to break up the subsoil in the second spit, enriching it with whatever can be obtained locally at a reasonable price: horse or cow manure (but not chicken or pig), preferably with much straw for heavy soils, spent mushroom compost, composted seaweed, or garden compost; peat; leaf-mould; wool shoddy—anything that will help to 'open up' the soil but still retain nutrients in solution and, if possible, contribute its own food-store. Well-weathered soot will help to make the tiny clay particles flocculate (stick together to make larger particles that drain and 'work' better), but any acidic soot or peat added should be balanced by a dressing of caustic lime, which has the extra advantage of absorbing water, swelling and helping to disintegrate hard lumps of clay during the winter.

All the above additives will help to create or increase that mysterious soil-condition known as 'humus'. Although inorganic fertilisers may bring quick short-term results, the long-term effect can be disastrous unless a reasonable amount of organic material is incorporated into the soil regularly. This produces or maintains the friable texture and encourages the development of beneficial bacteria in the soil; it also creates the right environment for a large earthworm population to consume decaying vegetable matter and help aerate the soil. To grow any crop really well, a high humus content is needed.

The soil should contain, in proper proportions, adequate food elements for the plant. There are some quite elaborate, but easy-to-use, soil-testing kits on the market; alternatively, you can send a sample of soil to the local agricultural advisory service for testing. If your plot is apparently the same throughout, mix small amounts of dry, weed-free topsoil from near the four corners and from the centre, and send it in a coffee or cocoa tin. However, if your plot appears to have variations of soil texture and colour, and some areas have differing weed-populations flourishing on them when allowed to go fallow, the samples from each area should be packed in separate tins and carefully labelled to indicate from which part each sample comes; the areas may require different treatments to bring them 'into balance'.

Most gardeners know, from the contents of 'balanced' fertilisers sold, that the nitrogen, phosphate, and potassium removed from the earth by successive crops have to be replaced, usually by the combination of sulphate of ammonia, superphosphate, and potassium sulphate. What the average gardener does not know is just what proportions of these ingredients will be most beneficial for his crops. This is why it is important to check the state of the soil before application, since it is most unwise to dose the land with individual fertilisers unless one is correcting a diagnosed deficiency.

Too little nitrogen causes gladiolus foliage to look yellowish and sickly. A quick corrective is a dressing with nitrate of ammonia, which releases nitrogen fast, but is also quickly washed down through the soil and out. However, too much nitrogen causes lush floppy foliage very susceptible to disease and also over-long flower-spikes, giving gappiness between blooms and a weak curving budtip.

Phosphate deficiency causes slow, inadequate root-development, resulting in smaller, less vigorous plants. Unfortunately, if this is not detected before the plants are growing above the ground, it is rather too late to take fully effective action. Steamed bone flour has more phosphate and less nitrogen content than bonemeal, but it should have been applied just before or at planting-time. Soil limed beyond the neutral point of pH 7 causes fixation of phosphate, thus making it less available to plants, which is the first reason for preferring to keep the gladiolus plot slightly acid. There is less danger from excess of phosphate, as about 75 per cent of what is added to the soil fails to dissolve in the soil-water and is never taken up by the plants.

Potassium deficiency reduces the health and vigour of the gladiolus and renders it more susceptible to disease, especially fungus diseases. It also weakens the whole metabolism of the plant, so that sugar production in the leaf is slowed down, resulting in less growth, a smaller leaf area presented to the sun, consequent less chlorophyll reaction, indeed a harmful descending spiral. In extreme cases, the leaf-tips die back early and the leaves turn yellow through chlorosis. Applied soon enough, potassium phosphate or (better still) a balanced fertiliser with additional potassium phosphate may save plants showing minor symptoms of this deficiency; but those looking really sick should be dug up and disposed of, as they could otherwise cause, in wet conditions, the start of damp-spot and mildew in the planting. As with phosphate, an excess of potassium is far less likely than an excess of nitrogen in the soil-water; but this does not mean that one should be heavy-handed in its use: sufficient of each to restore a correct balance is all that should be applied.

Magnesium deficiency can be caused by too much potassium and too little nitrogen being absorbed by the plant, even when there is a sufficient supply of magnesium in the soil, which is a rarity in British gardens. Be careful, therefore, to apply any extra potassium evenly at the calculated rate. The lack of magnesium interferes with the proper utilisation by the plant of existing phosphate and also seriously reduces the manufacture of chlorophyll. This sometimes causes gladiolus growers to think that their plants are suffering from *Fusarium oxysporum* (see p 54) when they are not, because the leaves tend to yellow between the veins through chlorosis. If you have apple trees in your own or a neighbouring garden, these will usually indicate which is the true cause of the yellowing, since on apple leaves it is accompanied by dark brown patches when magnesium deficiency is the cause and apple trees are not affected by *Fusarium.* Usually the deficiency in a garden or allotment is small enough for the gladiolus plant to grow normally; however, sufficient additional magnesium conveniently provided by a side dressing of Epsom salts will not only cause the foliage to become a deeper, healthier-looking green, but will also ensure that the flowers have stronger, cleaner-looking colours.

Calcium deficiency could occur in areas that were once heavily forested or for any other reason have acidic soil, but this should prove no problem if the ground is limed sufficiently to give a pH reading of 6·5 to 7.

Sulphur is more likely to be more than adequate in British

soils, especially in and around urban areas that have only recently stopped polluting the atmosphere with industrial and household smoke. Unfortunately, the rain in some areas still contains weak sulphuric acid, which causes blotching on dark-coloured gladiolus flowers, thus ruining their beauty. In the unlikely case of a sulphur deficiency, the gypsum in superphosphate will correct this.

There remain the so-called 'trace elements' which, although the proportion of each required in the total plant nutrients is minute, nevertheless are essential to the plant's well-being.

Iron deficiency is likely to occur on chalky soils or where the gardener has carelessly overlimed, again causing chlorosis and resulting in yellowing leaves. There may be plenty of iron in the soil, but the calcium carbonate precipitates it, so that plants cannot absorb it with soil-water. This is a second reason for preferring slightly acidic conditions for growing gladioli. Two remedial steps need to be taken in conjunction where iron deficiency is found to exist. The alkalinity needs to be counteracted by heavy dressings of acidic peat and leaf-mould, and growing plants need to be sprayed with iron chelate. Such chelates are often expensive, but fortunately small amounts make an enormous difference. The foliage spray will be partially absorbed through the leaves, thus supplying iron directly into the plant's system. That which is sprayed on to, or drips on to, the soil turns iron oxide atoms into part of a complex organic compound that *does* make the iron in the earth available to the plant, and natural chelates inside the plant then capture these iron atoms for the plant's specific uses. The pH figure for the plot should be no higher than 7·5 for satisfactory gladiolus growing.

Manganese deficiency is also associated with high alkalinity, though it occurs frequently in all types of soil. As the unoxidised form that is the only one of use to plants is more stable in acid than in neutral soil, this is a third argument for preferring a pH reading approaching 6·5. The symptoms are patches of yellow on the leaves. These occur because the protein synthesis has deteriorated, as has the plant's ability to 'breathe' through its leaf-stomata. Ensure that the drainage is good and that the pH reading is between 6·5 and 7, then such a deficiency should not occur. If it does, spraying with soluble manganese salts will remedy the plant's condition to some extent.

Boron, copper, zinc and molybdenum, though all essential, are required in such minute quantities and can be so injurious to plant-life if in excess that the layman would do well to avoid trying to correct any deficiency by chemical means. Russian comfrey, grown in a spare corner, stores in its leaves sufficient of these elements drawn from deep in the subsoil that its foliage composted with other vegetable matter will provide in safe amounts any additional 'trace elements' necessary.

A fourth reason for having the soil for gladioli rather on the acid side of neutral is to discourage *Fusarium oxysporum* (see p 54). Probably pH 6·5 is about as far from neutral as one should go, to avoid other drawbacks created by too acid a soil. Most British soils are in the range pH 6·4 to 6·9 and thus are ideal in this respect for growing gladioli.

PREPARATION

There are at least four methods of preparing a site for gladioli, each with its accompanying theory.

The first is definitely for the strong and vigorous only. It was employed in their student days by John and Robin Williams and enabled them, in 1963, virtually to 'sweep the board' at major English shows, even against commercial competition in the open classes. They won three major trophies and had the Grand Champion Spike at the British Gladiolus Society's national show; took the 'Tarbuck' and 'Amateur Gardening' trophies at Southport; won the Foremarke Cup and had the Champion Spike at the Royal Horticultural Society's Gladiolus Competition; and can hardly be said to have failed at Shrewsbury, where their 12ft × 6ft display of over 200 spikes was narrowly beaten by a commercial firm into second place.

The theory behind this method is that you want to grow gladioli, not weeds, and gladioli feed from a depth of 6in (150mm) downwards. Therefore, it is argued, the best soil with its aerobic bacteria and earthworms needs to be below the corms. To achieve this, the digging sequence is: remove the top spit (roughly 230mm) of a trench across the plot, then remove a second spit, break up the earth beneath to ensure good drainage and to incorporate whatever moisture-holding fertilising agent you are using to enrich your soil, and then turn the top spit of the next row upside-down into the existing trench (see Fig 5). This may sound like folly, since the second spit, which usually is mainly subsoil, is then placed above the first—something the gardener is warned against in most gardening manuals

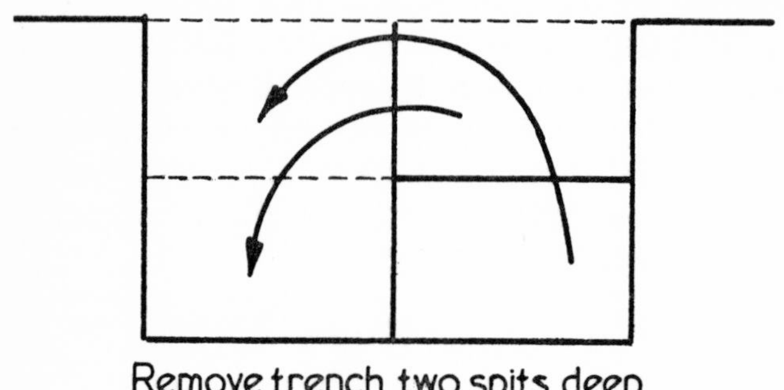

Fig 5 Williams's method of digging

and articles. However, what is achieved is a good depth of fertile soil into which the gladiolus roots may grow, the burying of weed seeds present to a depth at which they cannot germinate, and the possibility of increasing the depth of organic (as opposed to inert) soil permanently.

The method works well on light and medium soils, but is clearly unsuited to heavy clays, as the subsoil raised to the surface would be claggy and slippery in wet weather, and would harden quickly and crack in dry weather. On the other hand, in areas where the second spit is not largely subsoil or where the soil is light enough for aeration of the new top spit to be maintained, the depth of cultivation could be increased by this means and, after a few seasons of growing deep-rooting plants, shallow-rooting crops would be able to flourish again. If adopted, this method of preparation should be used in the autumn, to allow the earth to settle and the frost to work on the second spit now raised to the surface, which should preferably be left ridged.

The second method is that used by most exhibitors and conscientious gardeners, and so might be termed the 'normal'. Again the need to have nutrients where the roots will penetrate is recognised; but the increase in depth of fertility is made by incorporating the humus-making materials into the second spit and returning the soil to its original levels (see Fig 6). Double-digging or bastard-trenching is used, according to the time and energy available for the job. In double-digging, the first and second spits of the first row are removed and placed separately just beyond the far end of the plot. Then the top spit of the

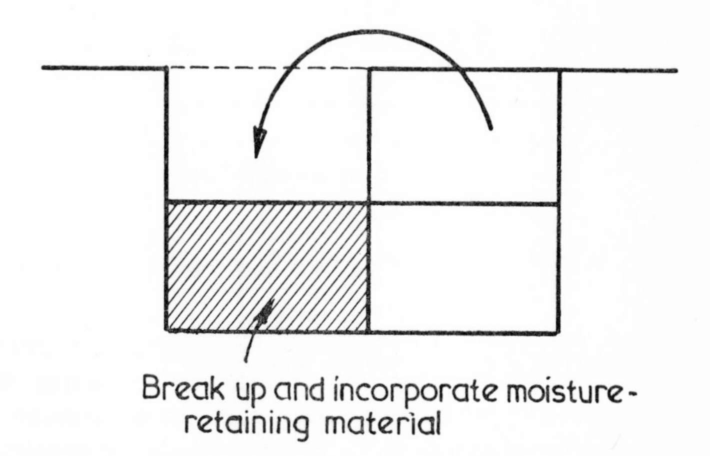

Fig 6 Normal method of digging

second row is removed and added to the other topsoil at the far end. The bottom of the first trench will be broken up with a fork, if this is thought necessary, and then the bulk water-retentive material will be thrown into the bottom. The second spit from the second row is turned on to this and some mixing done with the fork before the top spit of the third row is reversed above the second spit to fill in the first trench to slightly above normal surface level.

In bastard-trenching, the top spit only is taken out and removed to the far end. The second spit is then broken up as deeply as possible with a fork, the bulk material is thrown in, and the fork is used to mix it as well as possible into the second spit before the top spit of the second row is upturned on top of it. A further work- and back-saver is to divide the plot into two with a garden-line and, instead of barrowing removed soil to the far end, carry it on the spade to the other side of the line at the near end and make a ridge of it just beyond where the final digging will be done (see Fig 7). This halves the amount of earth that has to be transported and reduces considerably the distance it has to be carried.

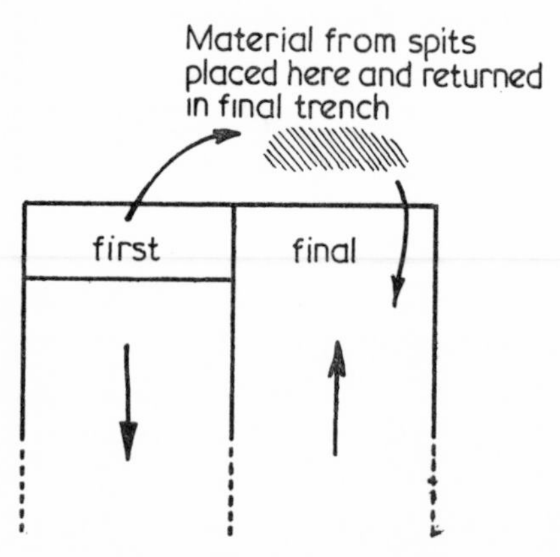

Fig 7 Divided plot digging

It should be emphasised that the above two methods of deep digging and heavily feeding the ground do not have to be resorted to to raise trophy-winning spikes from the smaller-flowered cultivars.

The third method—based on the theory that nature knows best and manages very well without man's interference—is particularly suited to older and to handicapped gardeners, as no digging is involved. Instead, a weedkiller such as sodium chlorate in solution is applied to the whole plot early in the autumn, so that the rains will wash it through the top two spits before planting time. The feeding is done by top-dressing with organic material and allowing the worms to mix this into the soil by ingesting, digesting and excreting it. As the worms also make drainage channels through the soil, the encouragement of a large earthworm population in this manner is obviously desirable. What is even more important, their activities help to create a good tilth and their castings bring rich well-structured material into the top few inches of the soil.

This is obviously good for surface-rooting crops, but is it so good for gladioli? The corms have still to be planted at the correct depth—though by making individual holes, not by taking out a deep furrow. Any goodness in the surface layer of soil will gradually be washed downwards towards the roots, but it will be washed farther unless there is a retentive layer to delay it.

The fourth method is a compromise that I find works quite well. If the soil has been deeply dug and enriched within the past five years, no autumn digging is needed, other than ridging the top spit for frost to act upon it if the soil is clayey. If no retentive layer has been put in place, then each area of the garden has to be dug until the whole garden has its 'underblanket'. For three or four seasons it is then possible to do merely light preparation in the spring. This consists of rototilling or forking the top 6in (150mm) as soon as the soil is dry enough, and then top-dressing it with compost and/or manure after planting. A fine top-tilth does not need to be produced where gladioli are to be planted, and this method enables fresh manure, when available, to be used, as it is not allowed to come into contact with the corms or their roots. The nutrients from fresh manure, leached through the soil by rain and later absorbed by the 'underblanket', are much more effective than those from old manure, already some 18in (460mm) deep, being washed even deeper before the roots can reach them.

Whatever method of soil preparation is adopted, it is essential to clean the site of all perennial weeds, either by hand-picking or poisoning, and to create conditions in which the fewest annual weed seeds will germinate. The first and second methods bury the original surface with its quota of seeds; the third method does not disturb the soil and bring dormant seeds to the surface, but relies upon shallow hoeing to deal with those weeds which do grow; the fourth method assumes that regular hoeing is going to be done anyway, to create a dust mulch and because some weeds are bound to grow, and will involve later mulching to suppress weed-germination and growth during the season.

CHOOSING CORMS

Do not leave your choice of stock until planting time is almost upon you. Try to visit shows, parks, nurseries and garden centres during the summer, to note the names of cultivars you would like to grow. There is no substitute for seeing the actual flower in bloom. Write for catalogues early in the autumn and place your order early—about November, not in the New Year.

If you are able to handle corms before purchase, remember you will get best results from a young heavy corm. A tiny basal plate (root-scar) indicates a corm grown straight from a cormlet without flowering; you will rarely find these available in large sizes. A small basal plate indicates a relatively young corm; choose this. A high-crowned corm that feels heavy for its diameter is both young and has a good 'food-store'; choose this in preference to flatter ones. The general rule is that the heavier a corm—provided it is still young—the better the results from any given cultivar under the available conditions of growth. However, some cultivars perform excellently from medium-sized corms, and primulinus-type gladioli rarely make very large corms.

It should be noted that American and Canadian top-size corms are termed No 1 and No 2 sizes, but these have diameters of 1·4in and 1·2in only respectively (equivalent to 103mm and 97mm circumference). Thus North American stock is always smaller than the Dutch stock; the Dutch top sizes are 12–14cm and 10–12cm. However, a prize-winning spike can be grown from an American No 2 corm, given good culture. Anything below this in size is likely to do better in its second year. On

Page 33 (above) A coring tool of this type can replace a trowel on light or friable medium soils and save much back-bending; *(below)* Corms may be stored in slatted trays or ones having a wire mesh base. These are being held in cold storage for July planting in Italy.

Page 34 (left) Each corm should have a handful of sharp sand laced with fungicide placed beneath and above it; *(right)* There are many 'knock-down' cupboards available that may be assembled at home on a landing or in an alcove to provide adequate storage space. The linen bags ensure good air-circulation.

receipt of parcels, open all bags of gladiolus corms and allow them to 'breathe'.

Any stock saved from the previous year should be carefully inspected, and diseased or suspect corms discarded. A count of the flowering-size corms of each cultivar should be made, to allow a plan of the intended planting to be drawn. Spare space could be allocated to the growing-on of small corms and cormlets; if there is insufficient space to plant all the corms available, then it is better to give the excess to friends rather than overcrowd the plot.

PLANNING AHEAD

One useful job can be done during winter evenings: the planning in detail of the sequence in which batches of corms are to be planted. This depends upon the purpose(s) in mind. For a long display in the garden, early-flowering cultivars will be planted first, followed by mid-season, followed by late. However, do not plant late-flowering cultivars after early May in the South of England, as there will be no time for the new corm to mature before they have to be lifted. Avoid late-flowering gladioli in the North, unless your garden gets the benefit of the Gulf Stream on the west coast of Scotland. In order to have the maximum number of spikes in bloom around show dates, the late-flowering will be planted first, then the mid-season, and the early-flowering last. Though it is often said that gladiolus cultivars bloom about a fortnight later in the north than in the south, this is a very vague average. Proximity to the east or west coast, altitude, darkness of soil, a sheltered or exposed site, whether it is in a frost-pocket—a wide variety of factors influences the time between planting and blooming. One learns by personal experience and from records—though a good shortcut for novices is to tap the fund of expertise in a local gladiolus society or group. Membership is not confined to national members; any interested grower may join and gain much in friendship as well as 'wrinkles'. It also offers another opportunity to inspect gladioli in bloom before ordering new ones.

Ready for showing Here are some pointers for getting gladioli into bloom on given show dates in England generally:

1 The number of days required to bloom varies from cultivar to cultivar and is reckoned from the planting time in already warm soil during the first half of April.
2 Very Early (VE) means less than 100 days *in the south*; Early (E) 101–110 days; Mid-season (M) 111–120, Late-Mid (LM) 121–130; Late (L) 131–140; and Very Late (VL) over 140 days.
3 Corms planted later than mid-April will bloom in slightly fewer days. As blooming-time is controlled by photo-periodicity (length of daylight), delaying planting by one week will make only a few days' difference in the flowering-time of full-sized corms.
4 Calculate the number of days from planting until the cultivar has six to eight fresh blooms open on it. This gives the number of days before the show date that the corms should be planted. Next season, divide your corms of each cultivar into three batches of at least six each. Plant the first a fortnight early; the second on the calculated date, and the third a fortnight later. This should give sufficient allowance for seasonal variation and ensure that *some* of the spikes will be show-worthy on the show day.

LABELS AND NOTES

For exhibition growing it is essential to keep your gladioli clearly labelled at all times; this is also desirable for all other growing of gladioli. However, if you object to having labels in your ornamental beds and borders, the names can merely be noted on a garden plan. Indeed, even when labels are used for rows, a clear record should be kept in a garden notebook, as birds, cats, dogs, not to mention small children, seem to delight in uprooting labels and leaving them in misleading places.

A typical notebook entry would read:

ROW 1: 12/4/75
Aurora 12L, 20M, 5S

Spring Song 6L, 5M, 2S
Simplicity 6L, 6M, 10S
Oregon Rose 4L, 2M, 6S

Row 2 etc. (L = large corms, M = medium, S = small.) Any failures can then be noted and, if labels become displaced, their correct position can be quickly ascertained.

One handy type of label is the white plastic kind with a hole at one end. This can be kept on the outside of storage bags by threading the drawstring through the hole before tying the final bow; it can be affixed to a storage tray by using a drawing-pin, and it can be placed in the ground at the start of each batch bearing that name. Moreover, it will not suffer from dipping, and can be held while immersing a bag of corms into a chemical solution. However, the sun tends to discolour these plastic labels and, from exposure to the elements, they become brittle. For these reasons I prefer to keep the plastic label with the bag—especially when the cormlets contained in it are to be grown on—and to use a second, wooden label in the gladiolus row.

Small labels in the ground are not only likely to get lost or buried, they are scarcely legible without much stooping. Long lasting, easily readable labels can be made simply enough by sawing 2in × ¾in wood into 1ft lengths, with a V-point at one end. The pointed ends are then stood in a container in 5in of green wood preserver (not creosote, which could be harmful to plants); after prolonged soaking, allow the ends to dry. Then paint the remaining 7in with a flat white paint or undercoat, using this space for bold vertical lettering in thick pencil to identify each cultivar. Such labels may be used over and over again, being repainted for new lettering whenever a cultivar is discarded.

TWO-TIER SYSTEM

As gladioli draw their sustenance from 6in downwards below ground-level, they are an ideal subject for people with small gardens who wish to get the greatest returns from the soil under cultivation. Between the positions of the planted corms shallow-rooting, early-maturing salading may be sown: spring onions, cos lettuce, small beetroots, radishes, and short carrots. The early germination of many of these seeds will mark the rows even before the gladiolus tips appear, thus making between-row hoeing safer. After thinning out, these salad crops will mature and be harvested well before the gladioli are fully grown, and thus one can get two crops from each row, one to please the stomach and one to please the eye!

PRE-PLANTING DIPS

To reduce the likelihood of infection from the soil and the spread of infection from diseased corms showing no external symptoms, many growers dip all their stock before planting. Mercuric chloride is usually advocated, but this must be used with extreme care. It is highly toxic, so rubber gloves and a face-mask should be worn when making the solution and dipping the corms. Follow the manufacturer's instructions exactly. A much safer dip to use is soluble Captan. Packets of sachets can be bought, each sachet being sufficient to make 1gal of solution. Whatever remains after dipping can be retained to use as a fungicide spray on the growing plants later. The newer Benlate is an alternative.

An additional advantage of a pre-planting dip is that it wets the root-nodules and encourages quick growth immediately after the corms have been planted.

PLANTING

Gladioli should always be planted in an 'envelope' of sand, unless the soil is very sandy already. This means that a handful of sand is placed under the corm, the corm is pressed firmly into it, and a further handful of sand is poured over the corm. As an alternative to dipping, therefore, fungicide and insecticide dusts can be incorporated into the sand beforehand. Incidentally, sharp silver sand should be used, not coarse builders' sand, as the intention is to ensure that the corm is never too damp, both to avoid rotting during wet seasons and to enable clean dry lifting with the minimum of scattering of cormlets in the autumn.

Distance apart The distance apart that corms will be planted depends upon the type bought and the purpose for which the plants are being grown.

For exhibition work, the giant- and large-flowered gladioli need spacing 1ft (300mm) apart in all directions; the medium-sized can manage with 9in (230mm); while the small- and miniature-flowered and the primulinus hybrids need only 5–6in (130–150mm) spacing in the rows, but still require 6–9in (150–230mm) between rows for ease of cultivation, attention, and cutting. Show flowers are most conveniently grown in blocks of three or four rows, with a 2ft 6in access path between adjacent blocks.

For cut-flowers for the market or the house, the above distances in the rows may be reduced by half; but the spaces between rows should never be less than 6in and preferably as much as 9in, especially for the larger-flowered, otherwise difficulties will occur as the plants mature.

Garden decoration distances can be the same as for cut-flowers, or a little more generous if large-flowered gladioli are being incorporated into herbaceous borders.

Depth Full-sized corms (ie 1½in or 40mm diameter and above) should be placed so that they are covered by 4in (100mm) of heavy soil, 5in (125mm) of medium loam, or 5½–6in (140–150mm) of light soil (see Fig 8). Deeper planting will adversely affect the number of flowerbuds produced and will reduce the number of cormlets spawned. Shallower planting will result in more but smaller cormlets and should only be resorted to for rapid propagation of seedlings or expensive new introductions, as the large-foliaged plants will be more prone to heeling over from strong winds, with a consequent loosening of the feeding-roots from intimate contact with the soil particles. Moreover, in a very dry season there will be a greater risk of the plant having an inadequate moisture intake at the critical time of maximum growth, ie when the flower-spike shoots up rapidly.

Obviously some allowance has to be made for the smaller initial food-store of smaller-sized corms, but 1in less than the above figures is quite adequate for corms 1–1½in (25–40mm) in diameter; and 2in less will suit the smallest corms below 1in across. As these form larger corms, so the contractile roots will tend to pull them to the optimum depth in the soil.

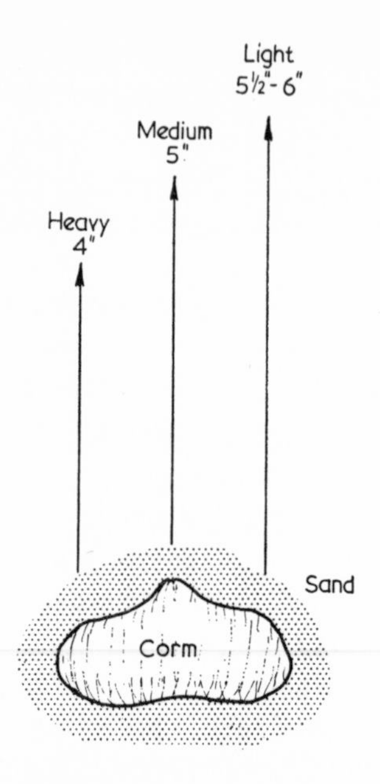

Fig 8 Depth of planting

Method If autumn preparation has been done or the no-digging method is being used, peg out a line, space the corms for that row at the correct intervals on one side of it, then with a trowel (not a dibber, which compacts the soil and can leave an air-pocket beneath the corm) make holes of the required depth on the other side of the line. These holes should be about 1¼in (30mm) deeper than the thickness of soil to cover the corm. Next pour a handful of sharp sand (laced with insecticide and fungicide, if possible) into each hole. Now press each corm firmly into this bed of sand. Pour a second handful of sand into each hole. Finally, rake back the removed soil, being careful not to allow any stones or hard lumps to go into the hole or lie immediately above the corm.

If, as is often the case, you had not decided what you were going to plant in time for autumn preparation, then a form of spring preparation usually gives quite satisfactory results. It is too late for deep digging, as the soil will not have sufficient time to settle adequately. Instead, take out a narrow trench about 10in (250mm) deep. Into the bottom of this place a 2in (50mm) layer of well-rotted manure, ripe garden compost, or well-soaked peat enriched with a commercial compost or balanced fertiliser. Rake in a 2in layer of garden soil and tread this down. Now place the handfuls of sand as little mounds at the correct intervals. Press your corms into these. Rake in more soil between the corms, drop the second handful of sand over each corm, and complete filling the trench.

As each row is completed, move the pegs of the garden line ready for the next row, allowing the width between rows that you have settled for, plus an extra inch for the leaf sheaves that will arise to form the 'neck' of each plant.

Immediately the corms have been planted, scatter some solid slug-killer across the plot and all round its perimeter, especially where there are grass paths or verges.

CARE OF PLANTS

Watering Within three weeks of beginning its growth, the gladiolus corm has begun the swelling that will create the new corm to be harvested; and, after about a month, the floral primordium has started to develop. To get the best from gladiolus plants, they must grow without a check right from the start. The ultimate spike is determined quite early in the growing cycle; too many people try to give flower-production a generous boost when it is too late to affect the number of buds.

If proper pre-planting preparation has been done, then the plant will grow steadily during the early stages. However, any stones or lumps of earth that impede the emergence of the leaf-sheaths should be removed as soon as noticed and, if the planting-period is a dry one, a good drenching of the gladiolus plot after about a week is desirable. Initially the filiform feeding roots have to pass through relatively dry sand; after this, they should get all the moisture they require from the enriched soil below. Never water lightly, by dampening only the top inch or two of soil, as this will cause the filiform roots to branch towards the surface, where they will be more easily dried out, instead of seeking downwards. Watering must be sufficient to soak right down to the roots, otherwise it will be more beneficial to the weeds than to the gladioli.

In prolonged dry spells, artificial watering will keep the plants alive, but they seem to 'stand still'. This is probably because mains water lacks the dissolved gases, particularly nitrogen, that rain water contains. It is desirable to have a rain-water butt beneath the fall-pipe of a greenhouse, sun-lounge, or conservatory, so that recourse to mains water is rare. Failing this, any means of aerating water in buckets or cans should be used. A tiny pinch of sulphate of ammonia to each 2gal is an alternative; but great care must be taken not to overdo this, as excess nitrogen leads to lush but disease-prone foliage, at the expense of the flower-spike. If a hose is being used, the sulphate of ammonia can be sprinkled very sparingly alongside the gladiolus rows.

The time to water copiously is when a thickening of the base of the leaf-sheaves can be felt between finger and thumb; this indicates that the flower-spike is starting its very rapid rise, and lack of moisture or nutrient during the following few weeks could be very detrimental.

Feeding However good the initial soil preparation, the larger gladioli will benefit from a few side-dressings hoed into the soil —preferably an organic substance, such as fish manure, or a balanced inorganic fertiliser. A little and often is better than sudden heavy dosages, but for practical purposes every three weeks from the time the plants are about 6in (150mm) high will suffice. The feeding that coincides nearest to the thickening of the plant base should have a little extra sulphate of potash added to strengthen the stem and a tiny amount (about half a teaspoonful) of Epsom salts to 2gal of water will add sufficient magnesium to accentuate the flower colour without ruining it.

Even when the spikes are being cut, the feeding should continue, as a living plant still remains to build next year's corm(s) and cormlets.

Primulinus hybrids and miniature-sized gladioli should not be boosted out of character. For these, a little bone-meal and ripe garden compost monthly, and that tiny dose of Epsom salts as the spikes develop, should be sufficient.

Spraying and dusting Once the leaf-sheaves appear above ground, it is never too early to spray. The whole object should be prevention rather than cure of trouble; moreover, chemical sprays can mar the beauty of flowers once they begin to open. Therefore your plants should spell death to all pests and fungus-spores for months before the blooms appear. Whenever you are using an insecticide in the garden, spray the gladioli as well, at no longer than fortnightly intervals. Should the weather be continually wet, use a puffer-pack of insecticide dust instead.

Warm moist atmospheres encourage the spread of fungus-spores. By June a fungicide should be incorporated into your spray and, if rain occurs often or mists—such as sea-frets—cover your plot, a fungicidal dust is called for. Never wait for trouble to appear—forestall it!

A helping hand During the growing season, gladioli rarely need much individual attention, but there are two things to watch for. Any plants not standing absolutely vertically should have the soil about their necks carefully drawn away by the fingers; then the plant should be very gently eased upright—a sharp or hard pull will snap the whole shoot from the corm. The soil should then be returned, pressed firmly around the neck, and extra soil raked up and heeled against the base. Try to disturb the roots as little as possible.

When the stems are thickening, showing that the spike is on its way, watch carefully for any plants that fail to let the tip out of the uppermost leaf. A few named cultivars are prone to trap the flower-tip and this causes the spike to bend as it grows unless assistance is promptly given. Just run the thumb-nail along to split the leaf-side and stroke the spike-tip upright.

The best helping hand of all is to uproot any diseased, yellowed or malformed plants as soon as they are spotted and destroy them away from the garden.

Staking Bamboo canes make the best stakes for gladioli; they are flexible enough to bend a little when strong winds strike the fan of foliage and are less likely to cause a flower-spike to snap. As 1ft of cane should be pushed into the ground, add this much to the ultimate height of the flowerhead when purchasing canes. This means that for the tallest exhibition gladioli 6ft lengths are needed, and proportionately shorter canes for the other gladioli. Primulinus hybrids and miniature-sized face-ups rarely need staking, except on very exposed sites. On many sites the small-flowered will stand up to wind and rain without support.

For cut-flowers that are going to be removed when first showing colour or when the first bloom is open, canes 1ft shorter are adequate. In commercial practice, two or three wires strained horizontally between posts set along the rows are often used for tying in, but this is only worthwhile in bulk-growing and there is a risk of damage to the flowerheads from the topmost taut wire.

For the actual ties, something very soft like wool causes the least damage, but take time in tying. Special handy garden ties about 200mm long, consisting of very thin wire through the centre of 5mm strong green paper are quicker to use and less conspicuous, but they are too short for the initial tying in of the foliage.

Staking, especially in exposed spots, should be done early. Insert the cane about 2in from the plant so as not to pierce or rock the corm. The foliage-fan of the gladiolus presents a considerable area to the wind. If tilted plants and loosening of the roots by rocking are to be avoided, then support needs to be given *before* the flower-spike appears. Once the spike breaks through, after a short while the curve of the tip (which always goes in the direction the blooms will face) will indicate whether the cane needs replacing to site it behind the opening flowers. If canes are originally placed to the north or east of the leaves, the majority will not require to be moved and therefore no feeding-roots will be suddenly damaged at a critical phase of the spike development.

When tying the actual flower-spike, remember that you are dealing with a rapidly elongating section and that over-tight tying can cause crooking. Make one reef-knot around the cane just loose enough to slide over any ridge above. Then tie a thumb-knot around the stem, just tight enough to hold the stem upright, and follow this with a bow or complete a reef-knot. Ties around close-packed buds should be very loose until they separate sufficiently for the original type of tie to be made round the stem between them. Watch that no calyces get caught and bent downwards by the tie as the spike grows. The bottom two ties are to support the foliage and the stem. Once ties are made between buds, if the loop around the cane is not pulled up by the additional growth, it should be slid above the level of the tie around the flowerhead. This will allow for three to four days' growth before further adjustment is necessary. Easing over the ridges of the joints in the bamboo is essential, so a quick inspection every second day or so is advisable at this stage. Keep the ties on the flowerhead down to the minimum required to keep the spike straight and upright.

Weeding and mulching No crop should be expected to compete with weeds and this is equally true of gladioli. Annual weeds are unlikely to be deep-rooted enough to compete with the gladiolus, but they do absorb nutrients and moisture from the soil before they can pass down to the gladiolus roots. The deeper-rooted perennial weeds, such as dock and creeping buttercup, not only compete with the gladioli for food but cause excessive evaporation during hot dry spells. Weeds are in fact a menace to the actual plant. Rain or heavy dew on weeds in contact with the base of the plant cause neck-rot and encourage slug-attack at soil level.

When the ground is moist, hand-weeding should be done close to the stems of the gladioli. When the ground is dry, hoeing should be done between the plants. If the corms have been planted with equal spacing in each row, then hoeing can be done not only between the rows, but also across the rows and diagonally through them too (see Fig 9).

The best way to reduce the chore of weeding is to mulch between and across the rows. This also conserves moisture that would otherwise be evaporated during hot weather. Mulching

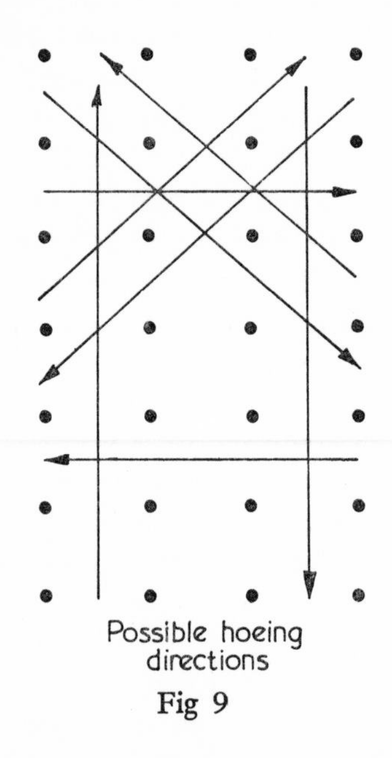

Fig 9

should therefore be done when the soil is wet, but no material that will rot or retain moisture should be in actual contact with any gladiolus stem. Mulching may be done with fresh manure (for large-flowered gladioli only); bracken, straw, peat (on alkaline soils only); sawdust (provided extra nitrogen is made available when this is eventually incorporated into the ground), shingle or small stone chippings. The most convenient and probably the cheapest way nowadays is to use long strips of tough black polythene; weight them down every few feet with large stones, house-bricks, or heavy clods of earth, to stop the wind from displacing them. Polythene has the added advantages of absorbing the sun's heat, channelling the rain into the plant-rows, causing condensation under it during hot weather, and being easily transportable and re-usable, as it may be rolled up at the end of each season, cleaned, and stored till the spring.

Cutting The art of cutting gladioli is to get a long 'handle' but still leave a minimum of four undamaged leaves to enable the plant to mature a satisfactory new corm. This is done by sliding a sharp knife down beside the stem, cutting vertically through the top leaves, then twisting the knife to slice diagonally through the stem, but checking the cut the moment the spike can be lifted clear, thus avoiding cutting too far into the opposite lower leaves. The 'handles' should, as soon as possible, be plunged into buckets of fresh water, unless the spikes are to be dried out before transportation (see p 70).

Lifting Harvesting the new corms and cormlets should not be delayed unduly, otherwise it may be impeded by wet cold weather. Each plant needs about six weeks from the removal of its flower-spike, in which to mature next year's corm satisfactorily. A note of the flowering date should be kept in the garden notebook, not only to calculate blooming-time for shows, but so that the earliest-flowering cultivars may be the first lifted and the rest harvested in the ideal sequence. Planting cultivars in rows or blocks according to their flowering-times facilitates systematic lifting. If foliage starts to turn brown early, then those plants should be lifted promptly, irrespective of the normal sequence.

The great mistake in trying to save gladioli for another year—a mistake perpetuated, unfortunately, by the 'expert advice' of so many general writers and broadcasters on horticulture, despite decades of campaigning against it by the specialist society—is to treat gladioli like onions, tying their leaves together in bundles and hanging them up in a greenhouse, shed, or garage. You may be lucky on occasion and collect clean healthy corms after they have dried, but it is actually courting disaster.

A far more satisfactory method is to take tomato-boxes, seed-trays, or other suitable receptacles on to the plot, with a liberal supply of dry newspaper. Line the base of any slatted box with a double layer of paper. As each cultivar (variety) is lifted, shake as much soil as possible from each corm without dislodging the cormlets. If the corms were planted in sand 'envelopes', as earlier advocated, this is where you will find one of the great benefits. Place all the corms of any one named cultivar into the box with their label. Have ready a second box or tray, paper-lined if necessary, into which the corms and their cormlets will finally be placed. This can be sectioned by troughs of newspaper if there are few plants of each cultivar.

Using sharp large scissors or pruning secateurs, cut the foliage off *as close to each corm as possible* before transferring the corm to the second tray. Inspect the foliage. Healthy foliage may be composted; anything diseased should be put into a bucket or similar container and later destroyed away from the garden or allotment. Tilt full trays slightly so that the sun and wind will do the maximum drying while you continue to work. The open secret of obtaining healthy corms is *fast drying*. This is why the foliage is cut off so soon. All long roots should be trimmed back at the same time and any wet outer leaf-husks stripped off. The less there is to dry, the quicker it will dry.

I am often asked, by those that are sceptical of these methods, 'What happens to all the goodness in the leaves? Shouldn't that be allowed to drain back into the corm?' The answer is that the tissue contains about 90 per cent water anyway, so the leaves will shrivel whether on the corm or off. Once the corm has been lifted from the soil, the work of the leaves and roots is finished. They are no longer of any use and are a mere hindrance. Worse than that, disease-spores flourish on dying leaves, especially in humid conditions, and can infect the corms if the foliage is left on. The minute increase in food-storage that could occur if the sap from the leaves did return to the corm is proportionately so negligible that it is not worth taking the risk of losing the corm. The clinching argument is that the method advocated here is that practised by the professional growers—and their livelihood depends upon having healthy corms to sell!

Once the corms are taken indoors, provided they are not cooked and dehydrated, fast drying may be achieved by placing trays in front of open coal fires, electric fires, or gas fires; stacking them near or above boilers; using electric fan-heaters or hair-dryers—in fact, employing any available means to get hot dry air circulating around and through them. Do not try to dry them in the humid conditions of a greenhouse or conservatory where growing plants are being frequently watered!

Cleaning Each tray of corms should be numbered in the order of drying, as there comes an optimum time, about a fortnight after drying, when the old corm-husk will pull away cleanly from the new basal plate. The corms should therefore be cleaned in sequence, removing the old corms and the dry dirt and debris. If the cormlets are to be kept for increasing your stock, the largest of these should be chosen now. Should you wish to give your cultivars a pre-storage fungicidal dip, now is the time to do it, followed by a repeat of the fast-drying procedure. Most amateurs, however, are content to dust their stock with a mixture of fungicide and insecticide powders.

Storage Examine your corms before storing and discard anything stone-hard, squashy, with a brown dampish basal plate, or pitted with black spots, especially where the leaf-husk rings are. Hard shiny black depressions are probably only the result of slug-damage. Using your thumbnail or a knife-edge, lever out the black surface. If this lifts easily to expose clean firm 'flesh' beneath, just dust the wound and retain the corm.

For storing small quantities of named cultivars, hung muslin or cotton bags are excellent, with the plastic label threaded on the draw-string. Larger quantities may be laid in slatted or meshed trays, but with no more than a double layer in each, so that there is plenty of air-circulation. Gladioli will winter quite safely in a wide variety of conditions obtainable in a normal home, provided the basic requirements are met. Storage must be dry, have occasional if not continuous air-circulation, preferably be dark (at least in early spring to prevent premature sprouting), and have a temperature between 2° and 10° C, ie remain above freezing-point but not exceed 50° F, the warmth of a mild winter's day. A frost-free garage is suitable, provided nightlights or a small paraffin lamp burn nearby during the severest weather. Similarly a loft may be used for storage if the heat from the house or the hot-water cistern keeps it above freezing-point throughout the winter. Any spare cupboard that is not in a regularly heated room can be converted to take trays and bags (see p 34). The cupboard door should be opened for a short time each week, unless air-holes are made in it. Even an old chest-of-drawers may be simply modified, with air-holes drilled in the front, or the drawers left slightly open.

If your ingenuity fails you or there is no cool place in the house, you can buy next season's sharp river-sand early in the autumn, pour it into boxes, tea-chests, or discarded or unused drawers, and bury layer after layer of corms in the dry sand. This will insulate your stock from cold, heat and light, and reduce to the absolute minimum the incidence of *Botrytis* through corms sweating, and the contamination of other corms by a diseased one.

Your stock should be inspected monthly throughout the storage period and any corms that show symptoms of disease removed. You may expect to lose about 5 per cent of your stock before re-planting. If you lose more, then your garden sanitation, drying, and storing methods need improving—always assuming you bought healthy stock to begin with; if you lose less, you are doing remarkably well and I would like to hear about your methods!

About a week before planting-time, bring your stock into warmer conditions and normal daylight to break dormancy; but do not encourage long shoots to grow before the corms are planted, as these rarely have the chance to grow upright.

Propagating from cormlets Cormlets may be planted in the

Page 51 (*above*) Propagating from cormlets: use deep boxes and make furrows filled with sharp sand. Press the peeled or cracked cormlets 2in down. Label and keep a record in a notebook; (*below*) Transporting exhibits: pegboard boxes for one layer of exhibition spikes only are easier to manhandle than the 'coffin' described on p 69; they can go inside an estate car or on a roof-rack.

Page 52 (left) By using good compost in boxes and liquid-feeding regularly, flowering-size corms can be produced from cormlets in one season; *(right)* As the pods begin to split, cut the stem and bring it indoors. Record the cross and flower number on a label attached to each stem.

open soil as with corms or, better still, in prepared boxes. The boxes should be at least 9in (225mm) deep, the bottom two-thirds filled with rich soil mixed with ripe compost or well-decayed manure. Then a 2in layer of John Innes potting compost should be pressed evenly over the top and the whole contents well soaked with a diluted disinfectant (Lysol or Jeyes' Fluid) a week or two before planting time, which can be earlier than for corms.

Skin off or, at the very least, crack the cormlet husks. Soak the cormlets for 12–24 hours in luke-warm water—for example, in a bowl above a hot-water tank or near a domestic boiler—keeping the cultivars separate in storage linen or muslin bags. With a suitably sized piece of wood, make 2in-deep troughs in the compost and fill them with sharp sand. Press the cormlets in so that they are about 2in deep, spacing them 2in apart along rows and keeping the rows at least 3in apart. Label each row and record it as it is completed, then top up the depression with sharp sand.

Give the cormlets the same seasonal treatment as for full-sized corms—except that, as the object is to produce corms as large as possible in one season, dilute liquid feeding should be given either weekly or fortnightly, immediately before rain or shortly after it. Never let the boxes get too dry. In prolonged hot weather, give them an occasional thorough watering.

Should cormlet plants try to flower, nip out the emerging bud-head. Allow the plants to continue growing as long as the foliage stays green. In late September or early October, you should lift healthy high-crowned corms $\frac{3}{4}$–$1\frac{1}{2}$in in diameter, all of which will give fine blooms for many years to come. To achieve this size, and reduce the number of fresh cormlets produced, add a final extra inch of sharp sand around the plants once the leaves are well developed.

Chapter 3

Diseases, Pests and Weeds

Some people are averse to growing gladioli because they believe it is difficult to maintain healthy stock from year to year. This is a fallacy. Provided three essentials are recognised—buy healthy stock initially, plant it correctly, and be scrupulous about garden and storage hygiene—your stock can not only be saved for replanting next year, but can also be increased.

Much damage has been done to the popular 'image' of the gladiolus by two things: mass-growers not being careful enough to see that their plants remain disease-free, and some hybridists concentrating much more upon the beauty and show potential of their introductions than their resistance to disease. Consequently, dozens of new cultivars come on to the market each year; but many last in commerce only a little while and can be a menace by introducing diseases into hitherto clean soil. Both dangers are now recognised and matters are improving with regard to the second, at least.

Always buy your new stock from a reputable source; that is, one that your national or regional gladiolus organisation recommends as having given its members satisfactory service for years past. Do not go for the 'cheap buy' simply because it is cheap—you will probably regret it and that will have been money wasted. Here are the things to guard against.

DISEASES

Probably the best-known disease is the premature yellowing of the foliage caused by *Fusarium oxysporum var. gladiolus*. There is no way of telling whether an otherwise sound-looking corm is affected, since the disease may be present internally, without any external clue—hence the emphasis upon buying only from reputable suppliers, who rogue their stocks regularly and take measures to eradicate this. Even if healthy corms are planted, the disease can enter through the plants' roots from already infected ground. Where possible, gladioli should be rotated on a five- to seven-year cycle. Slightly acid soil is preferable, because there is some evidence that a micro-organism which attacks *Fusarium oxysporum* thrives better in acid conditions and is therefore a useful natural ally to encourage.

In June or before, the leaves start turning yellow between the veins, which stand out greener. This is often called 'Fusarium Yellows', as it is a sure sign that the plant has this disease. All parts of the plant must be carefully dug up and burnt or otherwise destroyed clear of the soil. An examination of the roots will reveal that these have withered, as well as the leaves lacking chlorophyll. Be ruthless. Such plants cannot produce satisfactory flower-spikes; they can only produce trouble. If soil beneath the corm can be dug out, it should be disposed of, or sterilised.

Corms newly infected during the season's growth may not show the symptoms or show them so late that they are confused with the natural die-back and discoloration of foliage in early autumn. Therefore some corms may be cleaned and stored on the supposition that they are healthy. No dusting before and during storage can stop or decelerate this disease, as it develops an internal rot that turns the corm into a dark wrinkled lump. Such corms should be rogued out during periodic inspection of stored corms and destroyed. It is doubtful whether they will have contaminated adjacent corms, since the fungus normally enters by the roots; but any corms in contact with such a diseased one should be treated with a fungicidal dust in case spores or threads of the disease have lodged on them. Cool dry storage conditions discourage the development of the disease, but cannot cure it; this may lead to the false assumption that corms are all right at planting time because they have not shrivelled. This is probably the major way in which the disease is spread.

Every bit as serious, but more subject to control, is *Botrytis gladiolorum*. This enters the plant through small lesions in the leaves or leaf bases and will quickly travel to the corm, especially in damp conditions. This is why corms should be harvested about six weeks after flowering rather than later, with all the top foliage cut cleanly away with a knife, scissors, or secateurs, frequently dipped into a jar or bowl of disinfectant, and the corms dried as quickly as possible, ideally at 70–80° F. All the dried leaf-husks should be stripped off and the corms dried for a little longer before dusting and storing. During the growing period spores of this disease can alight on leaves, causing reddish-brown spots. If these spores release others to travel to the neck of the plant where it emerges from the soil, neck-rot is easily set up and this area turns brown—usually so weakening the plant that it will topple. Again strict garden hygiene must be practised: all diseased material being promptly removed from the site and burnt. Sometimes cutting off parts of leaves mildly affected may save a plant and avoid infestation, but this has to be done carefully and followed by a fungicidal spray or dusting. Infected corms harvested and stored will be subject to a soft rot, especially under very cool or cold conditions, and turn spongy with a whitish mould growing in and on them. Roguing and dusting must be done promptly, as this could infect a whole batch of stored corms.

Stromatinia dry rot can also cause neck-rot in corms, but it is found more in North America than in Britain. Small black pimples appear between the corm and soil-level. It has effects similar to *Botrytis* upon the growing plant.

If, towards flowering-time, plants suddenly collapse and die, this will usually be the result of *Sclerotinia gladioli*, a fungus similar or identical to *Stromatinia*, attacking the neck and the roots simultaneously. This fungus cannot live independently in the soil, but can survive there in plant debris from infected stock for up to five years. Even the new corms and cormlets will usually be infected.

Septoria gladioli begins as a leaf disease indicated by small round or square black spots. Young stock, including new seedlings, are particularly prone to this infection, especially during humid conditions. However, it can be fairly easily controlled by prompt weekly sprayings. Should it be left to run its course, the rotting fungi infect the leaf bases and then the corms, causing reddish-brown patches on the new corms, with lesions below. Cormlets can also become infected. Although the core of the corm is not initially diseased, the black sclerotia on the surface and the lesions, which have become black by normal cleaning time, will develop a hard rot that eventually turns the whole corm into the consistency of stone. All infected corms and cormlets should be ruthlessly discarded—even though it is claimed that spread within the corm can be retarded by high-temperature curing and cool dry storing. The danger lies in carrying the infection into next year's plantings, as the emerging leaves will carry the fungus-spores up from the corm or its husks and in wet conditions these spores ooze an orange secretion that can be splashed from plant to plant.

These four diseases—the main ones to attack gladioli the world over—can fortunately be not only contained, but prevented, by use of a benomyl fungicide. The sequence is: 4½ level teaspoonsful of powder to 1gal of water, or a 20gr sachet to 2gal of water. Dip the corms and cormlets for about half an hour and allow to dry before planting. Once the leaves have emerged, make a similar solution with the addition of some domestic detergent as a spreading and sticking agent. Spray all foliage at fortnightly intervals, increasing this to weekly intervals should any sign of disease appear. About two weeks before lifting, spray liberally into the heart of the plant.

To avoid the pre-planting soaking, it is possible to store your corms in a mixture containing this benomyl compound. Mix the powder with three times its volume of talcum powder. Pour it into a stout plastic bag or a large screw-top jar. Drop your corms in and carefully roll them around in the mixture, trying not to bruise them. All handling of this product should be done out-of-doors, wearing rubber gloves and a face-mask. The corms are then transferred to their storage bags or trays. These methods of treating gladioli with benomyl are much safer than the old mercuric chloride dip method, but it would be foolish to take unnecessary risks with any such fungicides—all are toxic to some degree. Store such mixtures in air-tight jars; do all dustings and dipping well away from food, cooking utensils, crockery etc; wash all exposed parts of the skin thoroughly afterwards, having first brushed or vacuumed all outer garments before entering the house.

Some growers believe that benomyl acts most effectively through the foliage, but could have a detrimental effect if it comes into contact with the roots, especially young roots just starting into growth. At the time of writing, there is no conclusive evidence of this. However, an alternative to avoid any inherent risk is to do the pre-planting dip in a Captan-based solution and use a different fungicide for storage dusting, even the old green sulphur. This should also be mixed into the sharp sand used at planting-time.

Other diseases also appear in parts of the USA and Canada. So that these are not spread to other areas, both state and federal phytosanitary (plant health) certificates are required with every importation of corms.

Curvularia trifolii is a fungus that attacks mainly cormlet plantings, causing 'damping-off' and showing as medium-sized shallow dark brown rot marks on the resultant corms. It may be minimised by hot water treatment, ie soaking the cormlets for thirty minutes in clean water at 55–56° C—just hot enough to kill the disease without killing the cormlet.

Certain viruses affect gladioli adversely and can be transmitted both by sap-sucking insects and by cutting instruments. When cutting flower-spikes and removing foliage at harvesting, a jar of disinfectant should be the constant resting-place of the cutting instrument when not in use.

Cucumber Mosaic Virus is mainly confined to North America, but occasionally appears in Europe, as the virus may be present in apparently healthy corms or cormlets. It causes white broken streaks in the foliage. Plants affected should be promptly rogued out. Often the resultant plant-size is affected; usually there are disfiguring colour-breaks on the petals, but the worst result could be subsequent infection of surrounding plants. There is no known cure.

Less serious are Tobacco Ring Spot and Bean Yellow Mosaic. The first causes chlorotic (light) spots on the foliage, usually disc-shaped. This reduces the efficiency of the food-making photosynthesis in the plant, but has little definitely known other adverse effect. However, plants showing the symptoms should be rogued out before other plants become infected. Bean Yellow Mosaic can cause streaking in the colour of the flowers, but is more harmful to beans than to gladioli. The leaf-mottling, mainly on younger leaves, is difficult to see unless the leaf is viewed against the sun, when it shows light and dark green angular patches, so early roguing is not easily accomplished.

Aster Yellows seems to be confined to the United States, and even there mainly to the north-west, because the carrier-insect is found mainly in that region. Plants affected turn straw-colour and die back. Those infected later produce spiralling flowerheads. Resultant new corms will not flower. At present this is fortunately a localised problem.

PESTS

The pests that attack gladioli can be divided into two main groups: those on or in the ground and those that fly.

Often the advice given for preparing ground for gladioli is to manure it and grow potatoes there the previous season. This is bad advice, because it can encourage potato eelworm, which also likes to feed on gladiolus corms.

If you have noticed wireworm at any time in your plot, remember that these will burrow through gladiolus corms, leaving them more susceptible to disease. To clear the area of both these pests, about a month before planting-time use a long dibber (one can easily be made from a broken fork or spade handle) to pierce holes 1ft deep and about 3ft apart in each direction throughout the plot. Then drop a few naphthalene flakes into each hole, fill it promptly with earth, and allow the soil-moisture to cause the whole area to be fumigated. A subsequent soaking with a disinfectant, such as Jeyes' Fluid at one tablespoonful to 1gal of water, applied at 2gal to a square yard, will not only accelerate the fumigation, but help to sterilise the soil. This should not be done later than two weeks before planting-time.

To the disinfectant may be added a soluble slug-poison, as slugs have a taste for gladiolus corms and roots, and for emergent leaves. Once planting has been commenced, solid slug-bait should be scattered liberally along the rows and around the planted area, especially close to grass paths, paving stones, and walls. You will probably be spared the larger marauders that I have had to contend with, such as browsing rabbits and even stray cattle, but if cats insist upon romping over your plot, then sprinklings of cat-pepper are indicated.

Among pests on the wing, some are perennial visitors to any plantings and must be guarded against in advance. The various aphides, such as greenfly and blackfly, not only damage the tissue of the foliage, but carry diseases and viruses with them. Therefore spraying with insecticide is a necessity from the time the leaves are a few inches high. A systemic insecticide is preferable, so that those that do feed on your plants are effectively poisoned, and also because a contact insecticide can be quite ineffective if your gladioli are invaded by thrips. *Taeniothrips simplex* is a minute winged creature, about $\frac{1}{12}$ in (2mm) long. Under a magnifying glass, it can be seen to be banded white and black round its body; to the naked eye it looks more like a tiny piece of black cotton. It crawls around on the gladiolus leaves, but does its egg-laying inside the leaf-sheaths, where no external poison usually reaches. It is here that the damage is done. The larvae, even though the parent may be dead, feed upon the succulent tissues of the inner sides of the leaves and cause irreparable harm. Apart from the presence of the adults externally, a silvery streakiness in the foliage is a sign that infestation has occurred. When the flowerspike does show through, the flowers usually will not open properly and the whole plant looks blasted. It is too late to take counter-measures once the damage has been done; spraying at least fortnightly (more often in wet weather) with a systemic insecticide is the only satisfactory defence. Some people manage to grow gladioli for a lifetime without ever seeing thrips; however, they are rarities. It is better to be safe than sorry!

Butterflies are delightful to watch, but not if they are settling on your gladioli. The common cabbage white is a particular menace. When the flowerhead is through, a neat round hole may be observed in one of the two top buds. That is the caterpillar's entrance—he is now inside, gorging himself upon the buds of those prize blooms you had hoped for. Later in the season, small dark globes will appear in the throat of open blooms. These are butterfly eggs, which, if not dealt with promptly, will result in the petals later having an unintended laciness. The systemic insecticide will deal with caterpillars too, though the initial hole in the tip may be made before the creature dies. A pair of tweezers should be used carefully to scrape the eggs from an open bloom into a jar before disposing of them.

WEED CONTROL

Gladioli should not be planted in ground that is already infested with deep-rooted perennial weeds, such as couch-grass, bindweed, wild rhubarb etc since, once the corms are planted, deep cultivation is impossible without damage or disturbance to the gladiolus roots.

However, occasionally neglected sites have to be utilised. In such cases the soil should be prepared early—at least six months in advance of planting-time—by poisoning the vegetation of the whole area with sodium chlorate, applied strictly according to the manufacturer's instructions. This chemical is highly flammable and should be treated with great caution. Do not use it during prolonged hot weather, since vegetation treated with it is subject to spontaneous combustion in such conditions. Dispose of used tins carefully. For the most effective results, apply it once in early September and again in early October. The winter rains will leach the residue down sufficiently not to harm corms planted at the normal times. Never Rototill couch-grass etc into the ground; you will only be multiplying your problems.

Annual weeds need to be dealt with promptly for two reasons:

they obviously compete for some of the nutrients in the soil and, whether after rain or a heavy dew, those near gladiolus spikes cause prolonged wet conditions where the spike is most vulnerable and can encourage neck-rot.

Frequent shallow hoeing and hand-weeding will, of course, keep the site weed-free, but this is very time-consuming and an unrewarding chore. If there is a large area to deal with, the use of a herbicide as a pre- and post-emergence spray is time-saving and more practical. It should be applied when the soil is moist and not likely to dry out quickly. The first spraying should be given after planting is completed. Subsequent spraying may be given after the gladiolus foliage has emerged, but care should be taken not to do this on warm days when droplets on the foliage can cause scorching. Stop using herbicide sprays once flower-tips begin to push through the foliage-fans.

One such spray that has proved satisfactory with gladioli is a chlorbufam-pyrazon compound, Alicep. This should be applied at the rate of 1lb to 25gal of water for a $\frac{1}{4}$ acre; so 1lb is more than sufficient for a whole season for anyone not growing gladioli commercially. Again, the manufacturer's instructions should be followed precisely. A stronger application could cause gladiolus leaves to come up twisted and be generally deleterious to the plants. This herbicide is most effective before the weeds have developed two true leaves, so it needs to be applied more than once soon after the soil has settled from planting. There should be no lasting ill-effects for succeeding crops, since it is warranted to break down harmlessly in the soil.

Another method of dealing with emergent weeds is to smother them by use of a mulch. This has the added benefit of conserving soil-moisture that would otherwise evaporate during the period when gladioli are at their most demanding for water. Various materials may be used: lawn-mowings (provided the grass is not in seed), peat, leaf-mould, straw etc, provided nothing decaying is allowed in contact with the gladiolus spikes.

The most effective material I have found to be black polythene sheeting of a sturdy thickness; with care it can be used year after year. Each prepared area of soil is covered by one or more lengths of black polythene, weighted down against the wind with house-bricks or large clods of earth at frequent intervals. Then, where the rows of corms are to be planted, an inch-wide strip is cut out with tailoring shears or large scissors to the distance needed to take all the corms (and cormlets, if propagating) of one cultivar. The flexibility of the polythene allows holes or a shallow trench to be trowelled out in the usual manner. Should labels become displaced, relabelling is easy with the open strips indicating the whereabouts of each variety. The great advantages of this method are: weeds are effectively smothered and cut off from sunlight; paths are relatively clean to walk upon, even after rain; the rain-water tends to run into the strips where the gladioli are, thus getting quickly down to their roots; the black polythene causes the soil to warm up more quickly and to retain its warmth, encouraging early and consistent growth; and moisture evaporating from the soil condenses on the underside of the polythene and returns, lessening the need for watering in hot weather. Because of these conditions, however, it is important to have poured soluble slug-killer over the area before laying the polythene and to sprinkle solid slug-bait inside the cut slits.

Chapter 4

Exhibiting and Judging

A great deal of fun can be had from growing gladioli for exhibition. This not only adds interest to gardening, but is one of the best ways of meeting and making friends with people from all walks of life. It is a mistake to think that this aspect of gladiolus-growing is 'a man's world'; one of the leading exhibitors in Britain is a woman, and women take part in the competitive classes of gladiolus shows from Moscow to California.

Moreover, if you want to find out what are the best gladioli to grow, then you must go to the shows. Only here can you see a widely representative range of the types, sizes and colours available. Here, too, you can glean information from others. There is usually someone present experienced enough to tell you what will grow well in your area and on your type of soil. Perhaps surprisingly, most exhibitors are only too willing to discuss cultivation, feeding, insecticides, fungicides, methods of transportation etc. I have been impressed during my twenty-odd years of gladiolus-growing by how helpful the 'old hands' are to novice exhibitors. Not only is there an absence of that sort of secrecy about the 'essential ingredient' that seems to characterise many growers of mammoth vegetables, but there is a real desire to encourage high standards of competition by passing on tips.

GROWING FOR EXHIBITION

Do not be deterred from showing by the thought that your spikes will not be at perfection on the date of the show; the same can be equally true for the other competitors. An element of chance is involved in exhibiting and your blooms could still be the best to appear on the showbench, especially at the smaller shows.

Timing The time from planting until the spike is at its optimum state for exhibition varies from season to season for any one cultivar. There is no substitute for experience, but even this cannot forecast the nature of the growing season and hence the speed with which any given cultivar will reach maturity. I have known April-planted 'Femina' vary from 103 days to 111 days. Planted in late March, it has taken 121 days. Planted in the second week of May, it has bloomed in as little as 75 days. The one fairly constant factor has been that in the South of England this cultivar opens in the third or fourth week of July if planted in April, but in the first or second week of August if planted in the first half of May. This has been true in Kent, Oxford, Somerset and Surrey. It can now be deduced that this is, comparatively speaking, a very early cultivar. The days are calculated to the first bloom on the first spike to open. Thereafter two or three days must be added, according to whether the cultivar opens two or three fresh blooms each day, as most show types will.

This will give a broad guideline: for late July shows in the south, plant 'Femina' during the first half of April; for shows in the first and second week of August, plant it in the first half of May, and for even later shows, plant it in late May. Such guidelines can be worked out for any cultivar. An experienced local grower will probably provide you with such information on many cultivars he has grown regularly.

However, the number of days to bloom still depends upon the weather conditions throughout the growing season. Having decided on the main show in which you wish to exhibit, make three plantings at fortnightly intervals, the second one timed to have blooms at their peak on this show date, if the season is normal; then find other shows on each side of that date into which earlier or later spikes may be entered.

If this seems to make planting too complicated, work out the approximate date of planting to 'hit' your main show and try planting a third of your corms an inch higher and a third an inch lower than the recommended depths for your type of soil. This will cause some staggering of the flowering-time, though not as much as would result from variation in planting dates.

Care during growth For show work greater care is needed for each individual plant. None of them should be allowed to suffer a check through water-shortage at any time during the season, so in hot dry weather provision has to be made for watering, whether by irrigation pipes or the plain chore of carrying buckets. Rain-water collected in barrels is better than tap-water and the opportunity can be taken to turn it into a liquid feed, by soaking a bag of old manure in it or adding a little proprietary fertiliser according to the manufacturer's instructions. Above all, do not skimp the watering. The larger-flowered gladioli in particular need a great deal of water in hot weather, as they have a large leaf area through which to transpire. All gladioli should be given extra water whenever a liquid feed has been applied—to wash it down to the roots and to ensure that the plants' water requirements are met without absorbing excessive nutrients.

Each time the plot is visited, especially after winds or heavy rain, check that all plants are as upright as possible. When the soil is fairly damp, it is quite easy to tilt up a plant that is leaning, provided that this is done gently, with the foot pressing the base of the plant in the same direction, so that there is no danger of snapping the foliage from the corm. Afterwards, rake a small mound of soil around the base to stabilise the plant and encourage the water to run off away from the neck of the gladiolus, which is vulnerable to disease attacks that involve rotting.

Once spikes are pushing their way through, see that none of the flowerhead tips fails to clear the foliage. Some cultivars are particularly loath to allow the tip to split the side of the central leaf; this results in the stem continuing to grow, but forming an arc because the top is trapped. In such cases, use the thumbnail to split the leaf-side, ease the tip out, then slide it gently through the closed hand, stroking it into an upright position. This may need to be repeated for two or three days if the curve in the flowerhead is a severe one, but it will straighten out if dealt with early.

Curves or kinks in the stem are less obliging. This is a case for 'putting the corsets on', as they say in the north of England. A cane is placed on the outside of the curve; the stem is tied in quite tightly, and the bottom of the cane eased gently into the soil quite close to the neck of the plant and in an upright position. A daily inspection needs to be made. The loops around the cane should be loose enough to be slid up as the stem grows, but often some retying of the spike is necessary before the curve can be completely straightened. If this does not work, there is the consoling thought that curved gladiolus spikes are in demand for house decoration and floral art.

Manipulation of buds Whilst inspecting and staking your gladioli, you can effectively influence the number of blooms which can be chosen for exhibition. During the final growing stage the manipulation of buds is very important. All buds, as they separate, should tilt out in the direction to which the tip of the spike tends to point. Some may attempt to go in other directions, even in reverse; this requires correction by careful persistent manipulation by gentle thumb pressure whilst holding the spike steady with the other hand.

Even when the correct facing of the flowers has thus been assured, the opening speed and the number of buds showing colour on show day can be marginally influenced by pressing forward each bud as it separates from the rest. When done skilfully, this can ensure a better 'taper' than the cultivar would normally give.

Cutting Cutting gladioli for shows has to be 'played by ear' on a day-to-day basis. The spike may be cut from the time the first bud in colour looks as though it will open next day. However, two opposing factors have to be taken into consideration. Spikes left on the plant will lengthen a bit more than those cut and put into water; they are also less of a nuisance left where they are than if brought indoors. On the other hand, whilst outside they run the risk of damage from high winds or heavy rain, or in sunny weather the colour on the lower blooms may fade. Use your own judgement here, according to weather conditions and available storage room.

Always cut early in the morning. At least four leaves should be left intact on the plant, so that the new corm matures adequately. Occasionally a low-set spike demands that the cut be made at ground level to get enough 'handle', in which case the whole plant should be uprooted, so that the corm does not just rot in the soil. Normally, however, the spike is held with one hand while the other slides the knife-blade down alongside it to slice through the upper, innermost leaves. When the blade is sufficiently low, angle it at 45° and make a slanting cut through the stem, taking care not to go farther and slice off the far leaves. A twist of the stem against the blade allows the stem to be pulled cleanly up through the foliage. Any leaves attached should be kept to give a 'finish' to the exhibit, although these are not essential and play no part in the pointing. Any badly damaged leaves should be cut cleanly off and disposed of, not left to decay on the plot.

Storing Cut spikes should be transferred to water as soon as possible, so that the cut surface does not dry out and harden. To keep spikes until the show day, it is essential to stand them perfectly upright in vases, jugs or cans, preferably with overhead light. Failing this last, then they should be placed as near to a good light source as possible, to prevent the tip bending in search of light. Spikes that are rather 'forward' should be kept in cool conditions, but a cold store around freezing-point often affects the colour. Spikes that need encouraging to open blooms quickly should be kept in warm conditions, such as a conservatory or greenhouse with a glass or clear plastic roof.

Fresh water should be given on each day of storage. If the period is longer than two days, on each second day about an inch of the stem should be sliced diagonally off the bottom, to ensure an adequate water intake. Sufficient water supply is an essential to retaining flower freshness as long as possible. Experiment has shown that the addition of 6oz of table sugar to 1gal of water used for storing and staging gladioli provides the necessary sucrose, and the addition of 0·1oz of 8-hydroxyquinoline citrate slows vascular stem blockage to enable the sucrose to reach the flowers. The life of the lowest flowers on the stem can be prolonged for about two days in this way. Probably the proprietary additives sold to flower arrangers contain similar ingredients; these would be a convenient way of supplying the correct amount of 8-HQC, which is only 600 parts per million.

Transportation Unless you specialise in exhibiting primulinus hybrids or the smaller-flowered non-primulinus, or the show happens to be close to your home, the transportation of blooms does become a problem. To get larger-flowered gladioli of 4ft or more in length safely to their destination, protective measures are needed.

The serious showman usually overcomes this by constructing one or more rectangular 'coffins' (see Fig 10), to suit his requirements. Typical dimensions are: 4ft 6in to 5ft long, 2ft wide and 12–15in deep. A firm framework is made from wood 1–1½in square and this is clad on the base, two long sides, and two ends with fixed pegboard. Pegboard is better than plywood, as it does not splinter, is less flexible, and—most important of all—allows air circulation. A screw-down lid is then made from pegboard and several more sheets are cut to fit into the box. These may be arranged in horizontal layers with spacing pieces, or made to slide vertically into guiding slots in deeper boxes (see Fig 10). The longer these sheets are, the more likely they are to bow slightly, so that the spacing—usually 5–6in—will need to be increased a little to prevent petals from being bruised.

These 'coffins' will give a lifetime of service and can withstand transportation on car roof-racks, by van or lorry, and in the guard's van of a train. If train journeys are likely, it is wise to fit four ball-castors to one end of the box, so that it can be upended and pushed along. A couple of handles will also be helpful.

Where only a few large-flowered blooms are to be transported or only a short distance has to be covered, florists' cardboard boxes can be used; the base and lid of each serving as a separate container. Make holes at one end so that the bare stems of the longer spikes can protrude and glue on a couple of slats of wood lengthwise as stiffeners to prevent the box from buckling.

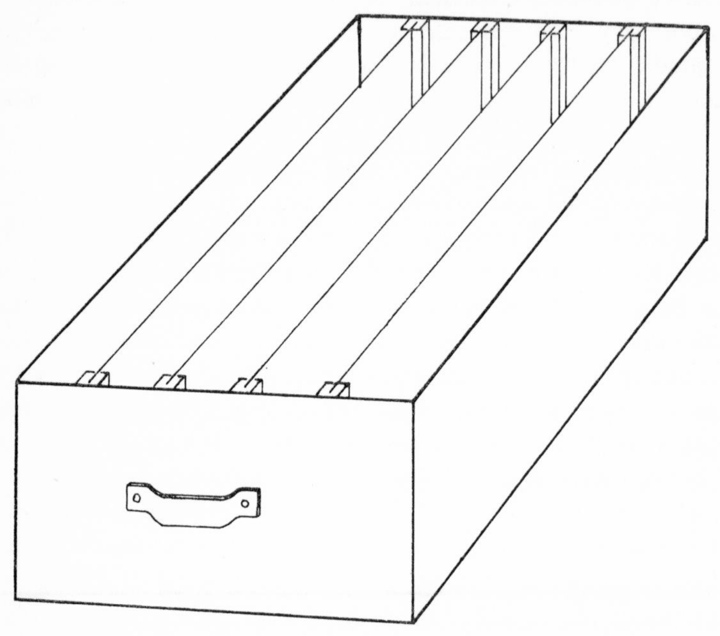

Fig 10 Multiple-tray 'coffin'

Packing If the journey to the show is a short one, the spikes may be kept in water until the last minute, then lifted out and the stems wiped before packing. On longer journeys the spikes are bound to dry out and the petals to flop, so it is wise to take the spikes out of water a couple of hours before packing to allow this to happen in advance. By packing when the petals are limp there is less danger of bruising or cracking from inadvertent pressure on them.

To pack the maximum number on one tray, place the gladioli head to tail alternately. A little preliminary experimentation with the spikes available should be done before any tying down starts. Often the smaller-flowered spikes can fit comfortably between the others, whereas two large-flowered spikes, though facing opposite directions, would still have their open blooms coinciding in the middle.

Short lengths of tape are then threaded through the pegboard holes to tie the spikes in position. For speed of untying, use bows. At any vulnerable points, and especially over the tip, place a piece of cottonwool under the tie before it is completed. On cardboard trays, Sellotape may be used to fix down, in which case cottonwool is needed at every fixing point and the sticky tape should have plenty of contact with the cardboard to ensure a firm grip.

Check list Any serious exhibitor is advised to make a check list of things to take to the show and to mark in a special way those that must be brought back. He should also indicate to the show secretary whether he wishes to collect his blooms after the show or is content to let the officials dispose of them as they choose. One important point to watch for in the smaller shows is whether the exhibitor is expected to provide his own vases.

Unpacking It is essential that dried-out spikes should have a long drink before being staged. Arrange, therefore, to arrive the previous evening or at least two and a half hours before judging commences. Unpack the tallest spikes first, since these need the maximum water-intake; cut an inch diagonally off the bottom of each stem, to expose the maximum amount of capillary tissue, and stand the spikes as upright as possible in buckets or other deep containers full of fresh water. Leave them until the petals have become turgid again. Now is the time to examine the spikes to see if any bottom blooms have 'gone over' and started to wilt. Put these aside, but do not discard them completely, as the removal of up to two lower flowers does not disqualify an exhibit. An excellent spike in all other respects could, though penalised for the removal of a wilted bottom bloom, still outpoint all others in its class. *It is a common misconception that a spike with a flower removed cannot win a prize*; it may do so, though it is less likely to, since the total bud count and overall balance of the spike are then also affected.

Staging While your spikes are recuperating, collect your exhibitor's cards from the show secretary's table and check that you have a correct card for each entry. Find the requisite number of vases of appropriate sizes—noting where the schedule calls for one spike to a vase and where three to a vase—and fill these with water to just above the halfway mark.

Never prepare your spikes on the exhibition benches. This is one of the few 'sins' that the novice has to learn quickly to avoid committing. The show committee does not want its cloth or paper covering on the exhibition benches soaked or stained, nor do other competitors want their exhibitors' cards to get wet. Make use of the uncovered wooden benches or trestle tables which are provided for the preparation of exhibits.

You will need an ample supply of paper—newspapers will do—and an old towel or piece of absorbent cloth. Select your spikes for each class and 'wodge' them firmly in the vase with the paper. A good judge will want to see the back of the exhibit and, if this is not easily visible from the other side of the show-bench or if tiered staging is used, vases will be picked up and turned round during the course of judging. Your spikes should therefore be set firmly so as not to move while this is going on, and not to twist round in any breeze blowing through the hall or marquee where the show is held.

A jug comes in handy for topping up the water-level close to the rim of the vase. Although nothing below the rim is taken into account in judging, exhibits always look more attractive from the point of view of public display if moss or some other suitable material is used as a 'finish' in the top of the vase. Carry the vase to the appropriate section of the exhibition benches and give it a final wipe down, especially under the bottom, before placing it gently in position.

Certain aspects of staging may affect placings in close competition. For instance, within a multi-spike exhibit, flowers that would detract from, rather than enhance, each other's colour should be kept apart—'a well-matched set' is the term often used of an attractive multi-spike exhibit. Provided the spikes are individually good, this should be one of the aims of the exhibitor.

However, it is far easier to achieve a well-matched set of one cultivar—since most of its spikes will be in bloom at the same time and they are all of the same build and colouring—and it should be noted that instructions to judges often specifically say 'Where the schedule does not specify otherwise, or there is an equality, preference for awards should be given to exhibits with the maximum variety of cultivars.'

Dressing The final dressing of the spike can also be important in close competition. Any insects should be wiped off carefully with cottonwool. If there are any dirt specks or marks caused by rain descending through a polluted atmosphere, these should be wiped away or minimised by gentle rubbing with clean dampened cottonwool.

The natural positioning of the blooms is governed by their opening. Usually the one above on the stem has its lowest petal in front of the bloom below. Occasionally, when a plant opens two or three flowers a day, the lowest petal of an upper flower may fall behind the top petals of the flower below. In such a case, with a finger or a pencil, the erring petal can be gently eased over the top, to create uniformity in the open flowerhead. This is better done when placing spikes into water on arrival, as once the petals are turgid there is more likelihood of cracking one that is moved. Any final manipulation with the thumb to press blooms and buds into the optimum position is also better carried out at this time, though a check should be made to see whether further manipulation is necessary before the exhibit is carried to the showbench. Wads of cottonwool may be used to 'persuade' the awkward bloom to sit correctly—but *remember to remove these* before judging begins.

Few gladioli, even among the recognised proven exhibition cultivars, have a natural perfection of taper. This is something that is often ignored, even among good showmen. The ideal gladiolus—and one is always judging with an ideal in mind, even if the flower in question only occasionally achieves it—should have a silhouette in which the outside edges follow lines that gradually taper inwards as they rise, to form an elongated triangle somewhat like Cleopatra's Needle on the Thames Embankment in London, but with a rather broader base (see Fig 11). Too often exhibitors are content to stage spikes which have a broad ribbon of colour that suddenly contracts into a narrow upright of buds, like a steeple perched upon a tower.

Nature usually needs a helping hand here and there. The lowest two blooms are usually broad enough for the base, but in

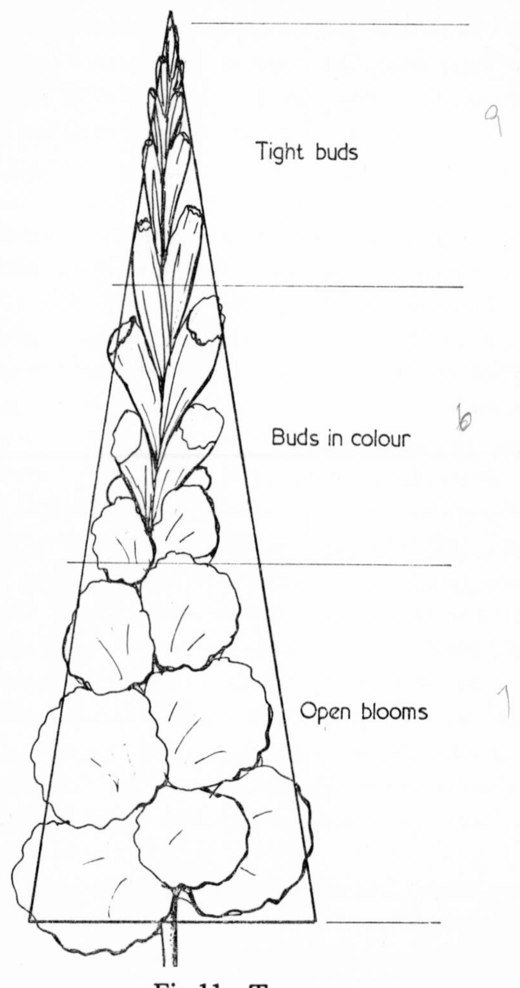

Fig 11 Taper

many cultivars the bottom one needs pressing gently upwards, by the thumb against the front of the calyx, from the day before it is about to open, otherwise it may sit too low. Provided it is not overdone so that the blooms no longer face forwards, the outside petals on the lowest two or four blooms can be gently brushed outwards with the back of the hand, if the width is inadequate.

The flowers above these need to be a little narrower across the front, so the opposite technique is used. By using the thumb and forefinger from the back, the calyces are squeezed slightly together in pairs and the outside petals are brushed gently forward. Care must be taken to avoid spoiling the facing of the flowers.

Few gladioli naturally give quite the right number of buds in colour to score the maximum points for this feature and to create a satisfactory tapering. Here the manipulation should begin early, while the spike is still on the plant or cut and stored indoors. As soon as the relevant buds separate, they should be gently thumbed forwards daily, until they show colour. Towards show time, if there is still a rather sharp narrowing above the open and half-open blooms, then by thumbing from the front the first two buds in colour may be persuaded to move farther outwards. Should your spike still be short of one or two buds in colour, then *very* careful easing to and fro of the unopened buds may cause them to show their petal tips, but the final position must look correct in relation to all the other open and unopened buds.

Any side-shoots (secondaries) should have been removed before cutting the spike; certainly they must not be left on an exhibit. However, the little piece of foliage immediately below the side-spike should be left on, so that the judge does not suspect that a faded or too low bottom bloom has been removed.

Height in vase The intention is to present a spike so that it appears to have a third of its height as bare stem, a third as fully open or partly open blooms, and a third as buds in colour and buds still completely green. This has been found independently in many countries to give the best appearance of balance. National societies have different views on presentation: in the USA and Canada milk cartons are used; the use of milk bottles for such purposes is illegal in Britain. Where vases are used, the British Gladiolus Society suggests that the 500, 400, and 300 size gladioli (see p 79) should be shown with the bottom of the lowest flower at least 6in (150mm), but no more than 15in (380mm), above the top rim of the vase, according to the overall length of the flowerhead. Wodging up with paper to compensate for a short handle is permissible, since nothing below the rim of the vase is judged, but care must be taken that the spike is still stable in the vase. The appearance for balance makes the onlooker take the vase into consideration. Excessively long stems obtained by uprooting plants and cutting immediately above the corm gain nothing and may lose points if this appearance of balance is upset when they are staged.

Accurate guidance cannot be given for the 200 and 100 size non-primulinus and the primulinus hybrid gladioli, since these vary so much in build. A spike with less than 4in of bare stem visible beneath the lowest flower may be adversely judged, but balance, as observed from directly in front of the exhibit, must be the guide to both the exhibitor and the judge.

Classes for gladioli In various countries classes may be arranged on different principles involving type, size, or colour. In North America, where there are many commercial growers, baskets of up to thirty or more spikes are sometimes asked for, and a class may be included specifically for garden clubs to make a joint entry. However, multi-spike classes are a relative rarity, the emphasis being upon single spikes in various colours and ranging over the five size-divisions that are becoming accepted generally in the gladiolus world. The only special classes for types other than the standard are the 'exotics' which usually cover the doubles, dragons, and fragrant gladioli. The Russians have a type called 'Hawks', but these have not yet been widely distributed and are little known outside the USSR.

Britain is almost alone in preserving the distinction between primulinus hybrids and non-primulinus types. It is a pity that this distinction is not more widely observed, because the primulinus hybrids are a distinct type much appreciated in both the garden and the house by those that prefer dainty flowers. Moreover, many of them seem to be more resistant to disease than do most of the non-primulinus types.

A primulinus hybrid is one that has in its ancestry the species *Gladiolus natalensis*, of which there are three yellow forms that were collected and named by three separate botanists: *G. primulinus* Baker, found in modern Tanzania; *G. nebulicola* Ingram, found close to the Victoria Falls; and *G. xanthus* Lewis, found in modern northern Zambia. The 'Maid of the Mist', introduced into England about 1902 and used in subsequent hybridising, was almost certainly *G. nebulicola*, as Mr Ingram always maintained; but, by one of those unfortunate twists of fate, it was thought to be *G. primulinus*, was termed so for long afterwards, and the whole race of its descendants now bears that name.

Early colour plates labelled *G. primulinus* and later drawings of 'Maid of the Mist' certainly show the same form (see Fig 12), with a deeply dipping hood-petal that completely protects the reproductive organs of the flower from the constant spray descending around the Victoria Falls. The modern hybrids from this maintain the hood-petal, in a less exaggerated form, so that it is possible to see into the throats without ducking almost to ground level. These modern hoods serve to protect the reproductive organs from rain and therefore increase the possibility of satisfactory seed-setting.

The term non-primulinus has had to be coined because the other summer-flowering garden gladioli were called 'grandiflorus', which is simply Latin for 'large-flowered'. With the development of smaller non-primulinus types, such as Len Butt's 'Ruffled Miniatures' in Canada, there came the absurdity of having classes labelled 'Miniature-flowered grandiflorus' and 'Small-flowered grandiflorus', which were as obviously self-contradictory as that traditional English expression 'Now then'!

In some smaller shows classes are made for 'Butterfly' gladioli. This is the trade-name for a race of mainly 300-size gladioli bred by Arie Hoek for the Dutch firm of Konijnenburg en Mark and which originally were characterised by large contrasting blotches on the two inner lip-petals. Unfortunately, several novelties added to this line are also termed 'Butterflies', though

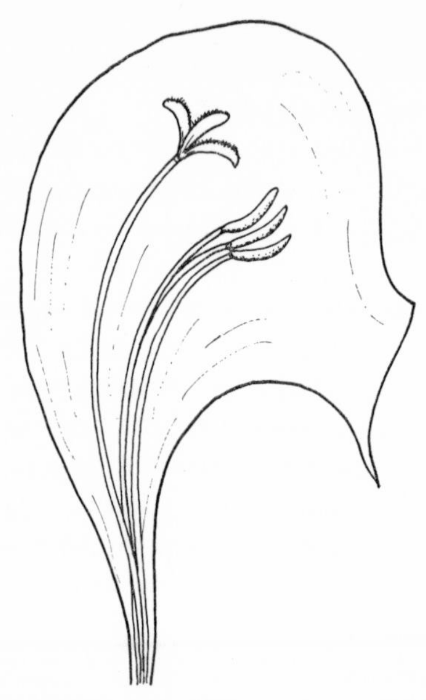

Fig 12 Protective hood-petal of *G. nebulicola*

there is nothing in their markings to justify the original delightful description of the type—that it looked as though a butterfly had alighted in the throat of each flower. These are not considered a separate type in British and North American shows of national significance; but, entered in their size classes, these cultivars are frequent winners.

Occasionally, in the largest shows in Britain, there is a class for face-ups or the newer 'Star' type, and no doubt other specific classes will be added as fresh types become more widely grown. Be sure to study each show schedule carefully to see what is valid in each class. In particular, a class that states: 'Any cultivar or cultivars' means precisely that; it can be won with good primulinus hybrids or a mixed entry of different types. Any show that aspires to any standing in the gladiolus world will also have classes (or at least one class) for new seedlings.

Classification As most shows have classes according to size and/or colour classification, and as most cataloguers use the code to indicate the size and colour class of the gladioli they sell, it is necessary to be familiar with, or at least have on hand for reference, what is rapidly becoming the accepted worldwide classification system. We are all indebted to the North American Gladiolus Council for inaugurating and revising this code, in which the first digit indicates the size, and the second and third the colour.

The British Gladiolus Society uses this system, but has taken it a step farther by adding a prefatory letter to indicate a special type, where it is recognised as being distinctive, starting with P as indicating a primulinus hybrid. Thus the code gives a handy partial description for any cultivar introduced. For example, P241 would mean 'having light pink primulinus-type flowers with conspicuous markings, ranging in size between 2½in and 3⅜in'.

The key to the code is as follows:

Sizes		*Diameter of flowers*
Miniature	100	Under 2½in
Small	200	2½–3⅜in
Medium	300	3½–4⅜in
Large	400	4½–5⅜in
Giant	500	Over 5⅜in

Colour	*Pale*	*Light*	*Medium*	*Deep*	*Other*
White	00				
Green[1]		02	04		
Yellow	10[2]	12	14	16	
Orange	20[3]	22	24	26	
Salmon	30	32	34	36[4]	
Pink	40	42	44	46	
Red	50[5]	52[6]	54	56	58 black red
Rose	60	62	64	66	68 black rose
Lavender	70	72	74	76	78 purple
Violet	80[7]	82	84	86	
Smokies	90 tan	92[8]	94	96	98 brown

[1] Chartreuse greens are classed with light greens (02); deep greens are classed with medium greens (04)
[2] Creams are also placed in this class (10)
[3] Buffs are also placed in this class (20)
[4] Scarlet-oranges are also placed in this class (36)
[5] As pale red is a form of pink, 50 will not be used; the number has been included to show that reds begin with 5 and that it has not been omitted in error
[6] Scarlet-reds are placed in this class (52)
[7] There are no true blues currently available. A medium or deep blue would be a form of the next lower Violet class, ie medium blue (82); deep blue (84)
[8] Smokies of all colour combinations from 92 to 96

Cultivars with conspicuous markings (CM) – blotches, clear darts, stripes, picotees, heavy stippling etc – different from the ground colour are assigned the odd-number digit one above: eg White with CM ends in 01.

The above 1973 NAGC revision is now the basis of classification for all cultivars internationally registered. The NAGC has undertaken responsibility for, and organised, an International Registry of the Gladiolus. It is hoped that all gladiolus raisers and distributors worldwide will co-operate.

JUDGING

Judges of the gladiolus should be aware that the effects of their decisions ultimately reach far beyond the show at which they are judging. Gladioli that frequently win at shows will be increasingly in demand by exhibitors, as well as acquiring a reputation for being good gladioli to grow. Indeed, the North American Gladiolus Council regularly publishes in its *Bulletin* lists of the 'Top Ten' in various categories and gives elaborate analyses of the winning cultivars in North American shows during the past summer. Such lists not only affect what people in the USA and Canada choose to grow, but help to determine what specialists in other countries decide to import from North America.

Similarly, the results of the Netherlands Gladiolus Society Colour Competition, published annually in *The Gladiolus*

Annual—the yearbook of the British Gladiolus Society—and elsewhere, help to decide what Dutch cultivars will be in demand outside Holland.

Whilst it is eminently desirable that judges of the gladiolus should keep up high standards, they should also remain receptive and flexible enough to recognise and do justice to new types. The Konijnenburg en Mark 'Butterfly' type was originally laughed at when first displayed in Holland; now these gladioli have enormous sales throughout Europe and in other parts of the world. Similarly, the primulinus hybrid type seems to be discriminated against outside Britain and perhaps New Zealand—where some were grown, bred and distributed by R. G. Wilson before his recent death—although it is also raised in Holland by P. Visser Cz. The Japanese used to have a penchant for the tinier-flowered gladioli and produced many face-ups.

The basic question to be faced is simply: 'Why do we grow gladioli?' It cannot be for their scent, since most summer-flowering gladioli are not perfumed and those that are can be described only as 'fragrant'. The answer must be that gladioli are grown for their beauty. However, when they reach the showbench, the exhibits are also being judged for the standard of their culture.

It is true that those aspects of the exhibit that can be counted or measured are far easier to judge than those that contribute towards the beauty of the flower. If it were simply a matter of counting and measuring, any ignoramus capable of these elementary procedures would be able to arrive at a first, second, and third in each class. Fortunately, the general consensus of opinion in the gladiolus world wishes due weight to be given to the beauty of the exhibit. In the score sheets that follow (pp 83 and 88–9) the NAGC points for colour clarity, saturation, harmony, and uniformity, added to those for Beauty and Appeal, and Taper, amount to 38 per cent. The BGS points for Flower Freshness, Beauty and Appeal, and Refinement and Balance of Flowerhead (which includes Taper) add up to 36 per cent for non-primulinus and 35 per cent for primulinus. The Flower Form is allowed five points by the NAGC; while the BGS allows four for non-primulinus, yet ten to the primulinus, where adherence to the correct form is very important. Although the NAGC treat Beauty and Appeal as purely subjective judgements, the BGS—whilst it recognises that some subjectivity is inevitable—tries to minimise these by giving detailed guidance.

It will be a sad day for the gladiolus, and flower-lovers generally, if ever the exhibits come to be judged purely upon 'mechanical' aspects because this is somehow imagined to be 'fairer'.

Another tricky aspect for the gladiolus judge is that among the larger-flowered types there are both formal and semi-formal placement of blooms. In the formal type, each adjacent bloom overlaps the other, but there is also no gap between one bloom above another on the same side of the stem. This presents a solid ribbon of colour that seems to mesmerise some judges and exhibitors, so that they incorrectly refer to this as the 'exhibition type'. A moment's thought should convince them that there are all sorts of exhibition types. In the semi-formal spike, the bloom above will still overlap the lower one, but the next one up on the same side is spaced so that there is an air-gap, although no stem should show from the front. There are many lovely gladioli that would be discriminated against if the formal were to be regarded as inherently better than the semi-formal, and the latter would probably never appear on the showbench—'Spring Song', for instance. The BGS makes a point of stating that 'a semi-formal cultivar should not be penalised for not having "solid ribbon" placement'. After all, it is readily accepted that in smaller sizes of cultivars—such as the so-called 'Ruffled Miniatures', which are usually 200 size—semi-formal placement and sometimes even more informal placement is natural to the cultivar. However, this is where the experienced judge's knowledge is essential: he needs to know whether he is looking at a semi-formal type or a formal type that has become overspaced through excessive feeding.

That this misconception about one type being inherently better than another is on the way out was amply proved at The British Gladiolus Society's Annual Show at Eastbourne in 1970, when 'Essex', a primulinus hybrid raised by Visser, was voted

Grand Champion Spike of the Show, thereby making gladiolus history. It also proved that a 'little 'un' can beat such cultivars as 'Landmark', 'Goliath', and 'Salmon Queen'—which were also in contention for the top honour—provided the judges look for the best exhibit *of its type* and do not exercise a built-in prejudice for any one type.

Not only must the judges know how to judge and point, but the exhibitors, too, need to know what is being looked for if they set their sights on major trophies or prizes. The BGS and NAGC score sheets and instructions to judges are therefore given here in full:

THE BRITISH GLADIOLUS SOCIETY
OFFICIAL JUDGING POINTS FOR GLADIOLI

	Non-Primulinus		*Primulinus*	
	Max	*Awarded*	*Max*	*Awarded*
1 Total Buds ... ... (6+6)	12	....	12	....
2 Open Flowers	8	....	8	....
3 Buds in Colour	3	....	3	....
4 Size of Flower	4	....	4	....
5 Spacing	4	....	6	....
6 Facing	6	....	4	....
7 Substance, Texture	8	....	6	....
8 Form of Flower	4	....	10	....
9 Calyx and Attachment	7	....	6	....
10 Flower Freshness	6	....	10	....
11 Beauty and Colour	20	....	15	....
12 Refinement and Balance of Flowerhead	10	....	10	....
13 Straightness and Strength/Form of Stem	8	....	6	....
	100		100	

BUDS: Six points for a basic number, but one point for each bud in excess to a maximum of six extra points.

Up to two flowers may be removed before judging. Deduct three points for the first and an additional five points for the second.

BASIC DATA

Class	*Diam in*			*Buds*	*Open*	*In Colour*
100	Under 2½	...	...	14	5	3
200	2½–3⅜	...	...	16	6	4
300	3½–4⅜	...	...	18	7	5
400	4½–5⅜	...	...	19	8	5
500	Over 5⅜	...	...	19	7	6
Score Points ...		...	...	6	8	3
Primulinus Hybrids		...	...	14	5	3
Score Points ...		...	...	6	8	3

Seedlings exhibited on the Showbench to be scored out of 100 according to type, then halve the result and add a further mark out of 50 for improvement upon existing cultivars and/or novelty.

INSTRUCTIONS TO JUDGES
(and Guidance to Exhibitors—Non-Primulinus)

1 *Total Buds:* This includes open flowers and buds showing colour as well as green buds. If a spike has the basic number for its classification, award six points, but add one point for each extra bud up to six (eg a 400 spike with twenty-five buds scores 12 points, the maximum). If the spike has fewer than the basic number, deduct one point from six for each bud short (eg a 400-size spike with only thirteen buds would score no points on this item).

2 *Open Flowers:* Eight points to be given to spikes having the number of flowers open for their classification as given in the basic data. Extra flowers open do *not* score any additional points. Deduct two points for each open flower short of the basic number. Flowers half or more open to be counted as fully open.

3 *Buds in Colour:* Deduct one point for each bud short of the appropriate number in the Basic Data.

4 *Size of Flower:* Assess on the lowest flower and penalise if the flower is below the classified size. Maximum points for the 300 to 500 sizes should only be given if the flowers are well up to the known size for the cultivar. Flowers on 100 and 200 sizes, if they maintain the general dainty and symmetrical appearance characteristic of these classes, should not be severely penalised for being slightly oversize (up to ¼in).

5 *Spacing:* The spacing between flowers should be even and regular. Penalise by one point for a too low bottom flower, even though it be characteristic of the cultivar. Formal cultivars should have flowers

slightly overlapping the one below, to present a continuous band of colour and it should be impossible to see the stem through the fully opened part of the flowerhead. Deduct for overcrowding, the excessive overlapping due to poor cultivation. A semi-formal cultivar should not be penalised for not having 'solid ribbon' placement. The stem should not be visible through the open blooms from a frontal viewpoint. It too should be penalised for overcrowding due to poor culture or excessive gappiness through over-boosting the flower-spike.

6 *Facing:* Flowers should face forward symmetrically. Any flower displaced to the wrong side of the stem should be severely penalised (deduct 3 points) and any flower (as it is common with the lowest flower on some cultivars) facing across the front of the stem should be lightly penalised (deduct 1 point).

7 *Substance and Texture:* These must be tested by feeling the petals very gently. The petals should be firm and resist displacement, without being too coarse. The physical quality of the petals may range from thin and pliable to thick and sturdy. The thickness should be even on all petals and throughout each petal, with no soft or cracked areas. Excessively thin, almost transparent, petals should be down-pointed.

8 *Form of Flowers:* This varies considerably among different cultivars and no particular form should be preferred over another, except that tubular form should be down-pointed in comparison with wider-open forms (face-ups excepted). There must be uniformity of flower-form, either single- or double-lipped, and any mixture that causes some flowers to appear inverted compared to the majority on the spike should be down-pointed according to the frequency of these inversions (deduct one point for each). Malformed or freak blooms should also be penalised.

9 *Calyx and Attachment:* Calyxes should be their natural colour and unmarked. Any split calyx should be penalised one point. Attachment should be tested by gently pressing the underside of a bloom upwards with a finger tip. The attachment should be firm enough to support the weight of the flower during moderate winds or heavy rain. Any attachment appearing weak should be penalised.

10 *Flower Freshness:* Penalise faded, browning, or curling flowers or any showing signs of limpness.

11 *Beauty and Colour:* Judges must not exercise any personal preference or dislike for any individual colour. Where the bloom is of one colour only, the intensity and any slight lightening or deepening of this colour in a bloom should be uniform throughout the whole open flowerhead. In a multi-coloured cultivar, there should be a pleasing harmony or contrast of colour. Unsharp edges to blotches, spades,

darts, picotees etc, should be penalised, but stippling characteristic of the cultivar should not. Smokies must be fairly assessed on a par with other cultivars. Unevenness of colour distribution and blending should be penalised. Shadings and markings should be uniform throughout the open flowerhead.

Any signs of a neutral petal-colour constitute a fault, as do rain- or spray-damage to the colouring. Deduct for abnormal flecking, for fading, for petals or parts thereof that have sported to a different colour, and for damage to or soiling of the petals. Judges are expected to use their aesthetic judgement to down-point any clash of colours that appears crude and unpleasant rather than striking and pleasant, and for anything that gives a muddy, dull appearance, as opposed to a clean, attractive one. Credit should be given for an overall crystalline glitter that enhances the flower's appearance.

12 *Refinement and Balance of Flowerhead:* There must be an overall pleasing, tapering appearance of open flowers, buds in colour, and green buds. Too many buds in relation to too few open flowers should be penalised, as should too many open with too few unopened buds. For relation of flowersize to flowerhead, a 300-size spike should have a 20in minimum flowerhead, a 400-size spike a 22in minimum, and a 500-size a 24in minimum flowerhead. For balance, the exhibit should appear to be roughly one-third stem, one-third a ribbon of colour, and one-third buds starting to show colour or completely green. Side-spikes must be removed, but enough evidence left to show that they were side-shoots and not flowers removed.

13 *Straightness to Tip and Strength:* Penalise for any kinks or curves in the stem, including the flowerhead, as viewed from the rear. The spike should be of sufficient strength to hold the flowerhead erect. Penalise only lightly for tips curving forward (½ to 1 point deduction, according to the degree).

OBLIGATORY DEDUCTIONS: Up to two flowers may be removed from the bottom of the flowerhead before judging. Deduct 3 points for the first and an additional 5 points for the second.

DISQUALIFICATION: Judges are obliged to disqualify all entries that are not according to schedule description for the class they are shown in. Additionally, judges should disqualify for any evidence of deliberate attempted deception, such as pinned-up or stuck-in blooms in place of the original flowers or any artificial means of holding flowers or part thereof in place. Appeals will be allowed against disqualification, but only on the terms stated in the schedule. A removed tip-bud (presumably damaged or accidentally broken off) is not grounds for disqualification, but should lead to additional down-pointing under Item 1, Total buds.

Multi-spike classes: These should be particularly carefully checked to see that they comply with the schedule. Individual spikes will not be pointed, but an average impression of the above items should be recorded as for single spikes. Where the schedule does not specify otherwise, or there is an equality, preference for awards should be given to exhibits with the maximum variety of cultivars.

Primulinus Hybrids

For items 3, 7, 9, 10 and 11, exactly the same instructions apply as for the Non-Primulinus Gladioli.

1 *Total Buds:* As for non-primulinus, but bearing in mind that the basic Data for Primulinus require only 14 buds, ie 20 buds score 12 points, the maximum.

2 *Open Flowers:* P100, P200 and P300 are all eligible, unless otherwise stated in the schedule. However, anything 4½in or above in diameter is uncharacteristic of the type and should be marked NAS, unless in a class calling for flowers of this description. Over-sized flowers for the cultivar should not be encouraged and gain no extra points, but they should be down-pointed only if the general dainty appearance of the type has been sacrificed.

5 *Spacing:* Flowers should be placed in a step-ladder fashion (ie as the feet would be placed in climbing a ladder) and spaced so that no flower overlaps another, but almost touches it. Penalise very lightly for slight overlapping, which is common to most cultivars of this type. Spacing should be even and regular. Penalise heavily for crowding or excessive gappiness.

6 *Facing:* Flowers should face forwards, but a slight outward facing is acceptable and should only be penalised if it becomes excessive.

8 *Form of Flower:* This is the most important characteristic of the type. The most pronounced feature is the hooding of the top centre petal to protect the pistil and stamens. Ideally this hood should be virtually horizontal and certainly not more than 22½ degrees above the horizontal indicated by the wing petals, though it may turn up at the tip, provided protection from rain would still be complete for the pistil and stamens. The two outer upper wing petals should show a distinct gap between them when viewed along the top of the hood. Spikes with one or more flowers not showing these distinctive primulinus hybrid features should be heavily penalised. Edge-frilling is acceptable, but heavier pleating towards the centre as in the 'Ruffled Miniatures' is not a primulinus characteristic and might render the exhibit NAS.

12 *Refinement and Balance of Flowerhead:* Approximately one-

third of the total buds should be open for perfect balance and the size of individual flowers should be in a good proportion to the total length of spike. The total effect should be one of daintiness, lightness, and airiness. Side-spikes should be removed, but enough evidence left to show that they were side-spikes and not flowers removed.

13 *Form of Stem:* This should be slender but strong, having a wire-like flexibility to make it whippy and able to bend easily in high winds. The stem is thin at the top and the final few buds are not necessarily held erect. Any kinks or curving as viewed from the back should be penalised, but a graceful curve forward when viewed from the side should not; only definite crooking when viewed from the side should be down-pointed.

OBLIGATORY DEDUCTIONS, DISQUALIFICATIONS AND MULTI-SPIKE CLASSES—as for non-primulinus instructions.

NORTH AMERICAN GLADIOLUS COUNCIL SCORE SHEET

Basic Data	*Diam in*	*Buds*	*Open*	*Buds Color*	*Flowerhead*
100:	2½	15	5	4	Maximum 22in
200:	2½–3½	18	6	5	Maximum 26in
300:	3½–4½	19	7	5	
400:	4½–5½	20	8	6	
500:	5½–	19	7	5	

PENALTIES — DEDUCTIONS

Item

Basic Data—deduct 2 points for up to 1 inch over maximum and 3 points for each additional inch, in flowerhead of 100–200.

8 −1 each bud short

9 −2 each floret short

10 −1 to −4 Judge's opinion

16 −2 for every inch over or under

All other items Judge's opinion

SCORE

FLORET

Color

1 Clarity	5:	———
2 Saturation	5:	———
3 Harmony	5:	———
4 Uniformity of Color	5:	———
5 Beauty and Appeal	10:	———
Structure		
6 Form of Floret	5:	———
7 Substance and Texture	5:	———
	40:	

A spike may also be penalised up to 10 additional points for each of the following serious faults:

Crooking .... ———

Attachment .... ———

Condition .... ———

Health .... ———

Deformed Florets ...... ———

Adventitious buds ...... ———

Stem too short, weak ... ———

Irregular opening ...... ———

Total additional deductions .... ———

3-Spike and 5-Spike Judging

Average all of the spikes under the One Spike System and then award up to 5 points for the UNIFORMITY of the spikes.

Bonus for uniformity ... ———

Final average score ...... ———

Total Final Score 3-Spike or 5-Spike ... ———

SPIKE

Structure

8 Total Buds ... 4:———
9 Open Florets ... 7:———
10 Buds in color ... 4:———
11 Floret size ...... 5:———
12 Facing .... 5:———
13 Uniformity of Florets .... 5:———
14 Grooming ...... 5:———
—
35:

Balance

15 Of Floret to Flrhead ... 9:———
16 Florescence to Flrhead ... 8:———
17 Taper .... 8:———
—
25:

TOTAL .... ———

LESS additional deduct ...... ———

LESS Penalty 100–200 flowerhead ... ———

FINAL SCORE ONE SPIKE ———

Instructions for Using the NAGC score sheet

Item

1 Clarity—deduct for dullness, muddiness, flecking, smearing or vagrant color.

2 Saturation — deduct for feathering, peeling, unevenness, bleeding of blotches.

3 Harmony — deduct for objectionable throat or lip markings, bizarre or discordant blotches, color combinations displeasing to the eye.

4 Uniformity of Color—deduct for fading or intensification of color from one floret to the next, in whole or in part.

5 Beauty and Appeal—purely subjective, the impact of the color and/or the form on the judge. The judge should strive not to permit the age or the frequent appearance of an exhibited glad to cloud his appreciation of its beauty,

or the novelty of a new one to bedazzle him.

6 Formation of a Floret—consider cupping, hooding, clawing, excessive reflexing, folded petals, ragged effect.

7 Substance is evidence of lasting quality and resistance to handling. Texture is the physical surface quality of the petal.

9 Open Florets—a floret shall be considered open if it is ½ or more the diameter of the next lower fully open floret.

10 Buds in Color—penalise for too many as well as too few.

11 Size of Floret—first floret establishes the size. Penalise 300, 400, 500 for undersize but not for oversize. Penalise 100, 200 for oversize only, deduct 3 points minimum for ¼in infraction etc.

12 Deduct for improper facing stem showing through etc.

13 Uniformity—one or two lip petals uniformly. Florets should not show varying degrees of rotation. Florets should be properly shaped to conform with their arrangement on the flowerhead.

14 Grooming—the presentation of the spike on the show table. Penalise for removal of any portion of floret or calyx. Side shoots should be removed but penalise if shoot sheath is removed. Side shoots may be left in place on seedlings.

15 Balance of Floret to Flowerhead—deduct for gapping, crowding. Florets too large or small for the length of head. Florets and buds should have a gradual and uniform diminution of the spaces between, from the first floret to the tip.

16 Florescence — measured in inches through the half opened buds and should be 50 per cent of the total flowerhead.

17 Taper—should also include the green buds. From the half open buds to the tip there should be a gradual separation, lowering and movement of the buds to alternate sides. The bottom of the flowerhead should be rounded.

Points to watch Although the gladiolus world is relatively honest, there is occasionally somebody that tries to 'pull a fast one' to deceive the judges. The most common trick is to replace a damaged, dirtied or faded flower with a fresh one from another spike of the same cultivar. This is usually done by means of a pin, the head of which can be noticed in the back of the stem; people with more ingenuity than wisdom or honesty sometimes fix the flower from the front, folding the calyx back over the pin to obscure it. Any shaky bloom should be investigated. Another dodge is to use a fine strip of transparent sticky paper to hold up a flower that is dropping too low, or to insert a matchstick into the calyx to prop it up. The use of any artificial supports above the vase-rim is forbidden and will lead to the disqualification of the exhibit.

It is my personal opinion that matters should go farther than this and that anyone found guilty of attempted deliberate deception should have all his entries at that show disqualified, be blacklisted, and be suspended from showing in that society's shows during the following season. Dishonesty should not be tolerated by societies that are trying to bring what is best and praiseworthy before the public.

A point possibly needing clarification is the wording to be found in many show schedules, including that of the British Gladiolus Society's Annual Show: 'Exhibits must . . . have been grown and flowered in the open.' This has never been used as an argument against cutting gladioli and completing their opening indoors, which is common practice. The intention is to rule out the raising of plants under glass or translucent plastic, and the sheltering of spikes still on the plant by the use of boxes on stakes with a roll-up piece of material in the front to protect the blooms from sunlight as they open. In other words, spikes left on the plants to gain maximum development are expected to be left open to the elements.

The question of more than one member of a family entering competitive classes is a delicate one. I would suggest that where husband and wife, or father and son etc wish to enter competitive classes based upon good culture, then they should furnish the show committee with satisfactory evidence—such as a statement from an independent witness of good standing—that their personal show spikes have been grown by each individual's unaided efforts. Until this state of affairs is true, then a joint entry of one exhibit to a class should be made—not so much because of any known misconduct here, but because any suspicion that a person has an unfair advantage can lead to ill-feeling, and both individuals and societies are entitled to be protected against this possibility.

Judges also need to be well acquainted with as many new introductions as possible, in case some unscrupulous person tries to pass one off as his/her own seedling.

From all the above, it can be seen that the judge's role is a crucial one for any flower. It is to be hoped that the time may come when gladioli are judged only by nationally accredited judges of the gladiolus, though possibly this will never happen in the smallest local shows. Meanwhile, panels of nationally accredited judges are being built up; but, as the standards required from a judge are high, the standard of the examination to attain that status has to be equally high.

Chapter 5

Growing for Profit or Pleasure

This book is for the ordinary person interested in growing gladioli and it is no part of my intention to attempt to teach the commercial grower his business, which he knows far better than I do. However, before listing some of the cultivated varieties available for growing in your garden—whether for decoration or show purposes—some glimpses into the world-wide scene of commercial growing and distribution may be of interest.

COMMERCIAL GROWERS

For an industry, even a horticultural industry, to show profits these days, it has to be highly mechanised. Much that has been written in Chapter 2 with the amateur in mind would be out of place in the commercial raising of gladioli and bulbous subjects. Here they are grown in carefully spaced single rows to make all the mechanical processes of planting, spraying, hilling and digging easily manageable without damaging the plants. In arid areas of the world, permanent irrigation along the rows must also be provided. Even the cleaning, sizing and counting into bags of the corms is mostly done mechanically, so that the human contribution lies mainly in supervision and the elimination of diseased stock.

There are, of course, many stages between the 'all-done-by-hand' garden growing and the fully mechanised industry. In North America some fifty wholesalers and retailers of gladioli advertise regularly, but these range from firms dealing in all sorts of horticultural supplies to families who make a second income from cut-flowers, corms and often their own seedlings grown on a limited area around, or near, their house. In 1973 there were about 10,500 acres under gladioli in twenty-three states, producing some 283 million saleable spikes.

In France, more than 70 per cent of the acreage growing bulbs or corms is devoted to gladioli. West Germany grows relatively few, but is a great importer of cut-flowers, even more so since the firm of Pfitzer stopped originating gladioli. East Germany has developed state farms on which to grow its own gladioli, to save importing them. It began with Dutch originations, later supplemented by some of the reliable 'commercials' from North America—a pattern similar to that followed by Bulgaria, Czechoslovakia, Poland and other Iron Curtain countries. It is sometimes surprising how far north the gladiolus will grow, even the rather pampered summer-flowering tetraploids. The Latvian SSR has its own gladiolus growers, based on Riga, and in the Moscow region much good work is being done in creating Russian originations, while many of the better American cultivars are exhibited at the Moscow shows.

In Japan, a country so long appreciative of flowers and artistry with natural materials, the number of hectares devoted to the gladiolus multiplied dramatically during the 1960s and may by now be exceeding the area used for tulips, lilies and iris. In the 1950s Ito was already producing Japanese originations with delightful names, such as 'Beni-Kocho', which I have had the pleasure of growing. Now they have their own introductions from 100 size to 500 size.

In Australia, a cohesive national society is difficult to form because the climatic variations are so vast that gladioli come into flower at different times of the year in the various states. For example, in Western Australia the flowering season virtually coincides with that in Europe; the growing season is throughout their wetter, cooler winter, their summers being unsuitable. Although some of the great originators of the past—Errey Bros., Both, Horrex etc—are no longer operating, W. M. (Bill) Blanden is continuing the tradition of Australian-raised gladioli with a string of outstanding ones, such as 'Ripple' and 'Royal Brocade'. His introductions, along with the best seedlings of a few amateurs, win 70–80 per cent of the prizes at shows in Western and Southern Australia. The only completely independent gladiolus society is the South Australian Gladiolus Society, which publishes the *Australian Gladiolus Annual.* Incidentally, an old Australian cultivar making a belated comeback in Britain is 'Ethereal', which, though plain-petalled, might be worth further hybridising.

Little has come out of New Zealand into the wider gladiolus world since the days when Norman Burn of Christchurch was originating 'Salmon Perfection', 'Centennial Summer', 'My O'Lovely', 'Special Edition', 'Dignity', 'Fifth Avenue' etc. Recently the emphasis has been upon Mrs Joan Wright's hybridising for fragrant gladioli.

The European market is currently dominated by the Dutch growers. From what seems a relatively small area—running along the west coast of Holland from north of The Hague, through Noordwijk, Lisse, Hillegom, Bennebroek, Heemstede, and Haarlem to Alkmaar, with small outlying areas at Sint-Pancras and Breezand—both cut-flowers and corms are exported, the latter even farther afield than Europe. The Dutch have three things in their favour: the good soil or 'geest' with a controllable water-table, thanks to their dikes; a government that appreciates the economic importance of floriculture to this relatively small country (in 1970–1 gladiolus exports alone totalled over 42½ million Dutch florins); and a traditional expertise at growing with an emphasis upon preserving good health. Between 1960 and 1970 gladioli accounted for 10–12½ per cent of the foreign exchange earned by floriculture.

Some firms in Holland specialise in selling large-flowered plain-coloured cultivars to recognised markets. Red is the most popular, especially in Germany; pink and yellow are also in great demand, but Catholic countries buy mainly white (for weddings and funerals). Even Cuba was supplied with white gladioli from Holland until Dr Castro took over and there was no longer the money available to buy them.

At least one Dutch firm uses Italy to produce its 'out-of-season' blooms for northern Europe. Although 'Friendship' and 'White Friendship' are held in cold storage until July, before being sent to Italy in time to flower for Christmas, 'Hunting Song' (light vermilion) is not planted in Italy until December, where it is grown under perspex to produce flowers in April. This is somewhat akin to the breeding of special cultivars for growing in Florida during the winter and then trucking them north for sales across the USA and even into Canada, except that in the Dutch case the same cultivars are used as are grown in Holland in summer.

British originations that have entered commerce have depended very heavily upon being the products of strong family firms dealing in other horticultural produce. George Mair & Sons, of Prestwick, Ayrshire, managed to get a few of their cultivars known outside the British Isles; Kelways of Langport, Somerset, had a run with their own originations, especially the 'Langprims'; but the best known of all British introductions have been those of W. J. Unwin Ltd, of Histon, Cambridge, mainly the work of Frank W. Unwin, who concentrated upon the primulinus hybrids, and then began the 'Star' and 'Peacock' ranges.

Britain is currently the melting-pot of the gladiolus world, accepting gladly the originations of all gladiolus-growing countries; her amateur hybridists hopefully crossing parents from very varied sources, and her national gladiolus society gallantly trying to bridge the gap between the 'metric' and the 'feet and inches' countries.

CULTIVARS FOR YOUR GARDEN

Of the thousands of named gladioli in existence, this is a personal selection to guide the newcomer and the relative novice. These gladioli will not behave identically under different conditions—as the reports from the British Gladiolus Society's test gardens, ranging from Scotland to the Sussex coast, have frequently shown. However, the majority will give great satisfaction in any soil and climatic conditions that could fairly be considered reasonable for this summer-flowering range of hybrid plants. Ultimately one chooses one's own favourites as a result of experience in a given locality.

These are listed by colour and in size-groups, using the

following abbreviations: B—Konijnenburg & Mark 'Butterfly' type; P—Unwin Primulinus Hybrid type; VE—Very early; E—Early; M—Mid-season; LM—Late-mid; L—Late; VL—Very late; C—makes a good cut-flower for commerce or home decoration; F—especially suited to floral art and home decoration; S—given show treatment, makes a fine exhibition spike; AA—All-America selection. The three-digit numbers are the size and colour classification code explained on p 79.

WHITE: Giant- and large-flowered:

501 Athena (Palmer) M, light lemon blotch, S.
400 Cotton Blossom (Fischer) LM, C S. 6ft spikes. (Certificate of Commendation, BGS).
401 Dream Girl (Frazee) M, medium pink stitching to petal edges, C.
400 Eastern Star (Fischer) LM, heavily ruffled, C S.
400 Icicle (Baerman) M, heavily ruffled, slightly green throat, C.
401 Jennifer (de Jager) M, throat blotched, C.
401 Lipstick (Fischer) LM, small mid-red blotches, C F.
400 Luxury Lace (Fischer) LM, heavily ruffled, faint cream throat, C S (Award of Merit, BGS).
400 Morning Bride (Fischer) E, C S (Award of Merit, BGS).
500 Mount Everest (Baerman) E, 6ft tall, up to 10 open, C S.
400 Parsifal (Walker) L, 9 ruffled open, 6 in colour, of 25 buds. C S.
401 Shooting Star (Baerman) E, salmon-red blotch bordered light yellow, C S.
400 Silver Chalice (White) LM, ruffled, clean, C F.
400 Simplicity (Fischer) LM, fairly plain, good taper, C F S.
400 Snowcap (Fischer) M, semi-formal, C.
400 Snow Dust (Rich) M, heavy texture, ruffled, C S.
400 Super Star (Walker) E, best US show white 1972, C S.
500 The Swan (Fischer) LM, ruffled, needle-pointed, C S.
500 Olympus (Marshall) M, sport of AA 'Anniversary', huge, C F S.
500 Wedding Bells (Roberts) M.

500 White Cathedral (Fischer) LM, tall, opens 10, S.
500 White Friendship (Fischer) EM, propagates well, C.
501 White Titan (Shearer) M, cream blotch edged lavender, C S.
400 White Wonder (Cartmell) L, very tall, many open, S.
400 Alba Nova (K & M) M, pink shine, C.
401 Emotion (K & M) M, creamy in throat, C.
400 Vesta (K & M) M, pure white, C.
400 Wimbledon (Visser) M, white sport of Antares, C.
400 Maria Goretti (Dutch) E, for a cheap garden flower.
400 White Tower (Dutch) L, ditto, but with possible show potential.

WHITE: Medium-flowered:

300 Bridal Wreath (Klein) E.
300 Chantilly Lace (Johnson) M, attractive decorative, C F.
300 Escalator (Roberts) M, tall, silvery-white, good opener, C S.
300 Marjorie Ann (Griesbach) LM, outstanding, reliable, S.
300 Rainier (Baerman) LM, old now, but still usually good, C S.
300 Silent Snow (Fischer) M, cream in throat, semi-formal, straight, vigorous, C F S.
300 Silver Bells (Roberts) LM, silvery-white, good opener, short handle, straight, C S.
300 Silver Wonderland (Griesbach) M, opens well, ruffled, graceful, C F S.
300 Snowdrop (Fischer) M, AA, slightly short, C F.
300 Wedding Bouquet (Baerman) M, rather old, still usually good, C F.
B300 Blondine (K & M) M, ruffled ivory with white throat, C F.
300 Sappho (Visser) M, pure white, C F.
300 Daintiness (Butt) E, creamy white, old but still good, C F.
301 Amapole (Visser) M, white with red fleck, C F.
301 Polar Beauty (Visser) M, ruffled white, cream throat, C F.

301 Bleeding Heart (Baerman) M, with red blotch, rather old, C F.
301 Blue Eyes (Fischer) M, AA, blue-violet blotches, C F.
301 Blue Flash (Baerman) M, alternative to the one above, C F.
301 Lucky Star (Wright–Gladanthera) LM, *Acidanthera* markings, for breeding for fragrance; has slight fragrance.
301 Repartee (Baerman) E, blotched red, edged yellow, C F S.
301 Second Love (Pierce) M, small blotch, good budcount, C F S.
B301 Daisy (K & M) M, creamy-white, large yellow blotches, C F.
B301 Dreamcastle (K & M) M, ivory-white, edged pink, dark carmine blotches, C.
B301 Little Doll (K & M) M, ivory-white, striking purple blotch, C F S.

WHITE: Small- and miniature-flowered:
201 Baby Blue Eyes (Fischer) LM, short, blue-violet blotch, F.
200 Cygnet (Larus) M, ruffled and good opener, C F S.
100 Half Note (Vawter) M, neat 'tiny tot', F.
201 Melissa (Fischer) E, ruffled and needle-pointed, pale yellow blotch, C F.
200 Mighty Mite (Larus) E, a winner for a decade, C F S.
201 Phantom Bantam (Pierce) E, speckly light purple blotch, F.
101 Safari (Roberts) M, large light purple blotch, good opener, ruffled and formally placed. Striking 'tiny tot', F S.
100 Tara Lee (Larus) E, getting old, still a winner for its size, F S.
200 White Lace, E, old but still quite good, ruffled, C F.
P201 Isis (DeBe) E, ivory-white, yellow throat, small and neat, F.

GREEN: Giant- and large-flowered:
404 Blarney (Arenius) M, mid-green, C F.

402 Green Bay (Frazee) L, tall, semi-formal, ruffled chartreuse, C F S.
502 Green Giant (Baerman), M, light green decorative, C F.
402 Green Goddess (Arenius) M, light green show-winner, C S.
402 Green Ice (Barker) M, one of the earliest pale greens introduced; still wins in Europe, C F S.
402 Green Willow (Walker) M, stretchy show-spike, C S.
402 Lemon Lime (Arenius) M, fades from green to yellow, but quite a good show record earlier, C S.
404 Forest Glade (Butt) M, mid-green, good opener, S.

GREEN: Medium-flowered:

305 Envy (Griesbach) M, mid-green with blotch, unusual, C F S.
303 Green Woodpecker (K & M) E, yellow-green with dark red throatmark. Used to be very dependable; now only carefully selected stock performs up to standard. Cheap. Good for breeding, C S(?).
304 Irish Spring (Turk) E, probably the greenest to date, C F.
302 Oasis (Roberts) E, ruffled lemon-chartreuse, good opener, C F.
B303 Armstrong (K & M) M, greenish-white, lemon-green blotch, C.
B303 Daily Sketch (K & M) M, greenish-white, yellow blotch and carmine stripes. Old now, but still good decorative, C.
B303 Impromptu (K & M) M, greenish-white, cherry blotch, C F.
B305 Prelude (K & M) M, green, large white blotches, pale purple stripes, C F.
302 Turandot (K & M) M, creamy-green, frilled, C F.
303 Corsage (Visser) M, green-yellow, flecked scarlet, C F.
302 Greenwich (Visser) M, ruffled greenish-yellow, C F.

GREEN: Small- and miniature-flowered:

203 Green Bird (Visser) M, ivory flushed green, yellow throat, C F S.

Page 101 (*right*) A prize-winning spike of 'Angel Eyes'; (*below*) Six good modern large-flowered exhibition cultivars: 'Green Goddess', 'Deep Velvet', 'Simplicity', 'Miss America', 'Apricot Glow' and 'Persian Rose'.

Page 102 (left) A prize-winning arrangement making good use of driftwood at base; *(below)* Here the driftwood has almost succeeded in following the spike-lines.

203 Bambade (Pierce) M, ruffled chartreuse with red throat mark, wiry stems, supersedes this raiser's earlier Bambi, C F.
203 Chickadee (Gove) M, green buds opening to chartreuse. Short but sturdy, C F.
202 Coquette (Butt) VE, valued for earliness, C F.
202 Early Green (Chandler) VE, first-class for floral work, but stock scarce, C F.
202 Emerald Isle (Butt) M, will catch most shows, F S.
204 Green Dragon (Rich) VE, ruffled medium green, F S.
202 Green Jewel (Harriman) E, 6–7 heavily ruffled, needle-pointed, C F S.
202 Green Sheen (Pierce) L, ruffled chartreuse, faint rose stippling in throat, tall straight spikes, C F S.
202 Little Jade Green (Pruitt) M, pure pale colour, F.
202 Leprechaun (Eppig) M, apple-green with 6 frilled blooms, stock restricted, C F S.
204 Mint Julip (Rich), VE, one of the greats: 7–8 ruffled blooms on whippy stems, C F S.

The smaller pale- to mid-greens are among the most attractive gladioli to use in floral art or as a foil to other colours in the home. We still await a really good green primulinus. Meanwhile, 204 Touch of Irish (Euer) VE, with a beautiful ruffled form and up to 9 open fresh, though rather short, will prove a favourite if sufficient stock is propagated and distributed. For a striking multi-coloured primulinus based on green there is P203 Margaret (DeBe) E, lime, striped wine, with upper petal and throat dark lilac, C S.

CREAM: Giant- and large-flowered:
410 Classmate (Roberts) M, tall, straight, ruffled, opens 9–10, strong substance, C S.
410 Dairy Queen (Griesbach) L, good show record, C S.
411 Ivory and Gold (Roberts) E, ruffled, golden blotch, good decorative, C F.
510 Landmark (White) M, great show record, but dull in colour unless given a trace of extra magnesium, S.

410 Lady Bountiful (Fischer) M, quickly became outstanding show and decorative flower in America, C F S.

410 Merle Doty (Griesbach) M, has long been one of best decorative creams, C F.

510 Old Ivory (Baerman) L, a good giant for garden and cutting, C.

410 Pale Moon (Roberts) M, slightly below the best show standards, C.

410 Party Ruffles (Rich) M, old, but still good and beautiful, cheaper than many, for mass plantings, C F.

410 Peace River (Fischer) M, wide-open, waxy-textured, ruffled, strong substance, C F S.

411 Ruffled Lotus (Baerman) M, heavily ruffled, C F.

411 Sculptured Beauty (Griesbach) M, fine form, good substance, C F.

410 Albesca (K & M), relatively cheap, C.

CREAM: Medium-flowered:

310 Cream Topper (Lacey) M, now rather old, but there is a distinct shortage of good creams in this size.

311 Duet (Dildane) M, still among the best creams for show and decorative work, C F S.

311 Fresh (Rich) M, also oldish, but a good decorative, ruffled, yellow throat, opens 9, C F S.

CREAM: Small- and miniature-flowered:

B211 Ariette (K & M) M, smaller than the average 'Butterfly', C.

210 Crinkled Wrinkles (Pierce) E, heavily ruffled, C F.

210 Dew Drop (Adams) M, pure colour, 7 open, wavy petals, C F S.

211 Edgelite (Butt) M, picoteed pink, C F.

211 Domino (Larus) E, speckled violet in throat, C F.

P211 Helene (Visser) M, edged vermilion, lower petals primrose, C.

P211 Pegasus (Visser) M, edged amber, veined fuchsia, C S.

P211 Perseus (Visser) M, edged orange-red, raspberry flecks, C F.

P111 Bonnie (DeBe) E, each petal edged bright rose, F.
P211 Estella (DeBe) E, 6 open with red, yellow, and lime blotch, S.

YELLOW: Giant- and large-flowered:
414 Aurora (Fischer) L, one of the best mid-yellows ever seen, C S.
415 Banana Split (Frazee) M, ruffled, small red blotch, F S.
414 Candleglow (Fischer) M, waxy, opens 8–9, C F S.
412 Candlelight (Rich) L, light yellow for garden and show, C S.
412 Empire Yellow (Rich) M, buttery in throat, semi-formal, C F.
512 Encore (Fischer) L, one of the biggest, combines Landmark and Limelight, rugged, lowest flowers semi-formal, C F S.
413 Gilded Crown (Woods) M, blotched deeper, C S.
416 Gold Coin (Van Staalduine) M, really deep colour, but not a great opener (6–7), occasionally manages more, C F S.
416 Golden Fleece (Roberts) M, pure deep ruffled, 8–9 open, C F S.
516 Golden Gigantea (Baerman) VL, only for those that prefer size to beauty.
416 Golden Glove (Fischer) LM, deeper even than Gold Coin, F S.
416 Golden Harvest (Griesbach) E, good early decorative, 7–8, C F.
416 Golden Peach (Gove) LM, rich colouring, C F.
416 Golden Sceptre (Fischer) M, sport of Sunkist, reliable, C.
414 Greeley Centennial (Roberts) M, pure, 8–10 ruffled, lasts well when cut, C F S.
415 Jubilation (Griesbach) M, buff-yellow, picoteed salmon, strong texture and ruffling. Best for garden and house.
412 Junior Prom (Eppig) L, great for shows and decorative, C F S.

414 Lemon Ruffles (Rich) E, old but useful early blooming, C F.

414 Lemon Lustre (Wiseman) M, shy opener (about 7), but still wins at shows; satiny surface, moderate ruffling, C F S.

414 Limelight (Fischer) L, AA tall ruffled, opening up to 10; high budcount, long a winner and still is, C S.

416 Morning Sun (Fischer) LM, AA, 8–9 open, now old and careful selection of stock advised, F S.

412 Ruffled Prospector (Baumbach) M, light, well ruffled, C F.

413 Siren (Baerman) M, good commercial, blotched, C.

514 Smaragd (Preyde) M, robust Dutch yellow, cheap, C.

416 Solid Gold (Baerman) LM, really deep, getting old now, C F.

412 Stylist (Roberts) M, ruffled, waxy, 8–9 on chartreuse side of light yellow, beautiful, C F S.

416 Summer Frolic (Fischer) LM, a deep yellow from Aurora and Limelight, spikes straight and open together, recommended C.

416 True Yellow (Baerman) E, good early deep colour, C F.

416 T 590 (Turk) M, strong, stretchy, ruffled, good opener, C S.

415 Eldorado (K & M) M, buttercup with blood-red blotch, C.

413 Forgotten Dreams (K & M) M, light with pink picotee, C F.

413 Gertrude Pfitzer (K & M) L, pinkish shine, blotched deeper yellow, good bud-count and opens up to 10, cheap, C S.

414 Nova Lux (K & M) M, lower petals darker than upper, cheap for garden planting, but large corms in short supply, C.

416 Royal Gold (K & M) M, one of the cheapest deep yellows, C F.

416 Yellow Emperor (Preyde) L, many buds, but opens only 6; useful at end of season, C F.

YELLOW: Medium-flowered:

316 Bit o' Sunshine (Fischer) VE, one of the first in bloom, C F.

317 Chinese Lantern (Griesbach) M, apricot-orange edge to petals, ruffled, good substance, 8–10 open, recommended C F S.

314 Folksong (Roberts) M, deeper in centre, ruffled, waxy, particularly good for decoration, C F.

314 Greenwich (Visser) M, ruffled greenish yellow, C F.

314 Jade Ruffles (Johnson) M, 8 ruffled greenish yellow, C F S.

B317 Walt Disney (K & M) VE, deep red blotch; old now, select stock carefully, C F.

317 Yellow Jacket (Baerman) M, striking red blotch, also old now, C F.

B315 Antoinette (K & M) M, barium-yellow, purple darts, C F S.

B315 Distincto (K & M) M, strong scarlet blotch, C F.

B313 Fatima (K & M) M, scarlet blotch edged mimosa, C F.

B315 Madrilene (K & M) M, light apricot-yellow, red blotch edged yellow, most attractive, C F S.

B315 Medusa (K & M) M, edged pink and blotched orange, C F.

B315 Page Polka (K & M) M, buttercup, blotched green, C F.

B315 Science Fiction (K & M) M, blotched red, C F.

312 Sweet Fairy (K & M) M, opens 10–12 on straight spike, C S.

315 Bristol (Visser) M, bright yellow, flecked red, C F.

317 Blackpool (Visser) M, amber-yellow, flecked red, C F.

316 Golden Horn (Visser) M, deep yellow, quite cheap, C F.

315 Saturnus (Visser) M, ruffled with dark blood-red blotch, C F.

317 Noisy Bee (Vennard) M, bright mahogany blotches on deep yellow, 7 open of 23 buds, ruffled, C F S.

YELLOW: Small- and miniature-flowered:

217 Brightsides (Fischer) VE, AA, orange-red edging, virtually primulinus form, opens 6 or 7 out of 18, C F S.
212 First Chance (Pierce) E, light colour, needs coaxing to open many, C F (S?).
212 Fraulein (Fischer) M, opens only 6, but good clean colour, C F.
216 Golden Rosebud (Butt) E, old, relatively cheap, C F.
216 Goldilocks (Fischer) E, good deep colour, winner for many years, C F S.
216 Golden Elf (Pickell) LM, for late shows and display, C F S.
215 Kon-Tiki (Griesbach) M, medium with red blotch, C F.
214 Lazy Daze (Baerman) M, clean medium, C F S.
112 Mimi (Roberts) M, tiny, good decorative, F.
216 Nugget (Baerman) M, 7–8 frilled open, leader in this group, C F S.
117 Scout (Roberts) VE, opens 7 out of 18 waved flowers, F.
215 Statuette (Butt) E, old, cheap, but still going strong, C F.
215 Sun Frolic (Blanden) M, good Australian, C F S.
217 Sunset Sky (Balentine) VE, reliable garden and decorative, C.
213 The Imp (Baerman) E, red-blotched face-up, C F.
212 Towhead (Larus) M, long the leading pale yellow; do not overfeed, C F S.
212 Josie (DeBe) E, opens 8 ruffled, cleaner Statuette, C F S.
P214 Yellow Poppy (Visser) E, cheap pure yellow prim., C F.

BUFF: Giant- and large-flowered:

420 Adventure (Roberts) E, heavily ruffled, greenish throat, F S.
420 Amberlight (Jack) M, honey amber, golden throat, 8–10, C F S.
420 Apricot Delight (Fischer) L, suede-textured, needle-pointed, a little short on buds and opening, but beautiful, C F.

420 Apricot & Gold (Baerman) L, descriptive of its blending, C F.

420 Apricot Lustre (Rich) L, 8–9 peachy-buff, yellower in throat, one of the best for all purposes, C F S.

420 Aprigold (Butt) M, 8 frilled, sometimes opens later, C F S.

520 Bali Hai (Blanden) M, Australian giant with punning name, S.

520 Blonde Beauty (Tyndall) E, ruffled pinkish-buff, yellower in throat, huge, C S.

420 Happy Birthday (Fischer) LM, 10 lightly ruffled, C F S. Excels.

420 Honeycomb (Fischer) L, well ruffled, small red line on each lip, very beautiful, C F S.

420 Illuminator (Fischer) LM, mellow, subtle blending, exhibition placement, C F S.

420 Lamplighter (Baerman) LM, 8–9 open of 24 buds, C F S.

421 Silent Love (Fischer) LM, almost yellow, edged salmon-pink, extremely beautiful, F S.

BUFF: Medium-flowered:

320 Fashion (Fischer) LM, heavily ruffled, rather informal, pearly lustre, strong texture, opens from green buds, C F.

321 Queen of Hearts (Griesbach) M, blotched red, C F.

320 Sundown (Baerman) M, 7–8 open of 22 buds, heavy texture and ruffling, C F S.

BUFF: Small- and miniature-flowered:

220 Bit o' Honey (Fischer) LM, very lovely apricot-buff, C F S.

120 Buffette (Adams) VE, the only 'tiny' in this colour, F.

220 Novelette (Butt) M, now old and difficult to obtain.

220 Perky (Fischer) M, yellow with apricot-pink overtone, ruffled, heavy substance, waxy, very attractive, C F.

ORANGE: Giant- and large-flowered:

422 Klondyke (Fischer) LM, golden-buff, tinted pinkish-buff on upper petals, tall, intensely ruffled, real gold! C F S.

427 Autumn Glow (Fischer) M, deep gold suffused red at edges, somewhat informal, but beautiful, C F.

523 Autumn Sunset (Thayer) L, huge new light orange, red throat, C F S.

426 Bittersweet (Frazee) E, vigorous, 6–8 open, ruffled, heavy texture, waxy, C F S.

526 Coral Blaze (Shearer) M, coral orange, flecked gold, 8 open of 23 buds, C F S.

426 Coral Glow (Fischer) LM, opens 10–12 in exhibition placement, 22–24 buds, consistent, C F S.

425 Far West (Roberts) M, yellow centre bordered salmon, tall, ruffled, 7–8 open, C S.

526 Fireball (Larus) M, flame-orange, picoteed white, opens 7 plus, C S.

427 Gayway (Roberts) E, apricot-orange, yellow centre, 7 ruffled open, slightly short, but early, C.

426 Golden Pheasant (Dirk) LM, slightly short in budcount, but very attractive, C F.

424 Golden Rosette (Baerman) VL, old, still good, pure colour, C F S(?).

526 His Nibs (Pazderski) M, huge, deep coloured, C S.

527 Isle of Capri (Pazderski) M, salmon-orange with deeper orange-red blotch edged white, lightly ruffled, 7-10 open of up to 25 buds, C S.

524 Madam Chiang (Glines) M, good decorative, fair show, C F S.

426 Orange Chiffon (Larus) VL, AA, tall, 7–10 open, heavy substance, ruffled, formal placement, C S.

424 Orange Crush (Fischer) M, vibrantly 'luminous', ruffled, a shade informal, up to 9 open, C F S.

524 Orange Rocket (Rich) L, huge and tall, faint throat-marks, opens 7 plus, C S.

424 Orange Spire (Salman) LM, sturdy, healthy Dutch, C S.

427 Saxony (Baerman) M, deep, blotched, with show prospects, S.

422 Setting Sun (Baerman) LM, good light colour for garden, decoration, and show, C F S.
427 Sunkist (Fischer) M, strong orange with yellow lip-petals, formal placement, C F S.
422 Sunny Morn (Baerman) LM, old, but still good, C F S.
425 Toulouse Lautrec (K & M) E, coral-rose, buttercup yellow blotch with red mark, may be forced, C.
425 Happiness (K & M)—not to be confused with Baerman's 'Happiness'—salmon-orange with white blotch, E, C.
427 Orange Globe (K & M) M, coral-red with blood-red mark, C.
Beau Geste (Blom & Padding), Dutch colour champion in 1972, C S.

ORANGE: Medium-flowered:
323 Carioca (Squires) M, best American decorative and show, strong substance, ruffled, 7–8 open, orange-red throat-spot, C F S.
327 Chiquita (Eppig) L, excellent decorative, good show, C F S.
327 Coral Seas (Griesbach) E, ruffled, deep throat, C F.
325 Discovery (Roberts) M, ruffled, large chartreuse centre, C F.
324 Forever Yours (Fischer) LM, heavily ruffled, blend of light salmon and bright yellow to give apricot effect, C F.
327 Gypsy Dancer (Fischer) LM, old, but striking, C.
323 Ming Toy (Brush) E, blotched show and decorative, C F S.
322 Orange Beauty (Howell) E, consistently good for several years, clear, light colour, C F S.
B325 Roulette (K & M) M, blotched 'Butterfly' type, C F.
B327 Aladdin (K & M) M, red blotch on lemon base, C.
B327 Andalusie (K & M) M, orange-red, large yellow blotch, C F.
326 Thalia (K & M) M, plain orange-red, C.
B327 Claudette (Visser) M, carrot-orange, red spot on yellow, C S.

ORANGE: Small- and miniature-flowered:

P226 Orange Velvet (Visser) E, strong-textured primulinus, C F S.

P226 Alicide (Visser) E, orange-red, edged white, C F S.

P225 Guildhall (Visser) M, with canary-yellow blotch, C S.

P227 Tangerine (Visser) E, deep orange with red stripes, C F.

125 Buddy (Roberts) VE, widely flared top petal, lower petals clear chartreuse with deep orange midribs, 6–7 ruffled, C F S.

227 Fiesta (Griesbach) M, deep orange border to bright yellow centre, top petal salmon-orange, 7–8 ruffled open, C F S.

226 Foxfire (Roberts) M, outstanding scarlet-orange, 8–10 open, C F S.

224 Little Mo (Lake) E, best of the lighter oranges, C F S.

126 Orange Brilliant (Vawter) M, another delightful 'tiny', old now, gold midribs on red-orange, 7–8 ruffled open, C F S.

126 Scooter (Roberts) VE, slightly marked in throat, 7–8 open, C F S.

224 Sweet Melody (Baerman) E, apricot-orange with yellow throat, short spike, but 7–8 ruffled open, C F.

122 Toyland (Roberts) E, bright yellow centre, opens 8–9, heavily ruffled, F S.

SALMON: Giant- and large-flowered:

432 All Ruffles (Baerman) LM, good light decorative, C F.

435 Beauty Queen (Bissenden) M, cream throat, heavily ruffled, 8 open of 22–23 buds, C F.

532 Big Daddy (Ruppel) LM, AA, scores on size, both spikes and flowers being huge, but has little beauty appeal, S.

432 Brigadier (Howell) L, yellow throat with brown peppering, 11-12 open of 24 buds; old now, but still wins when grown well, C S.

532 Chinese Coral (Rich) M, best of the light salmons in America, C F S.

536 Conquest (Griesbach) M, aptly named, this deep salmon

shot rapidly up the popularity charts throughout the world, C F S.

437 Flying High (Larus) LM, white throat, tall, C S.

534 Frilled Champion (Baerman) M, small cream throat, 8–9 ruffled, of 20 buds, C F S.

534 Goliath (Baerman) LM, old now, but grown well still approaches its original 10 open on up to 26 buds, white throat-mark, C S.

430 Good News (Larus) M, pale apricot-salmon, C F S.

435 Heritage (Griesbach) M, white throat, top show and decorative, C F S.

434 Longfellow (Melk) M, clean colour, heavy substance, C.

436 Mountie (Melk) M, scarlet-orange, ruffled, very heavy substance, C F S.

537 Parade (Larus) M, up to 10 open of 25–27 buds, ruffled, heavy substance, cream throat, a champion all the way, C S.

436 Red Deer (Butt) M, sparkling salmon-red, many open, C F S.

436 Salmon Queen (Schrenk) M, creamy throat, good placement, frequent show-winner, C S.

433 Sister Fortuna (Griesbach) M, lower petals creamy white, 6–8 open, very ruffled and crimped, waxy, decorative, C F.

431 Summer Garden (K & M) M, pale healthy Dutch, C F.

435 Towering Queen (Calhoun) M, creamy white throat, opens 9–12 of 24–28 buds, C S.

436 Thunderbird (Henderson) LM, AA, erratic, can give good spikes of 8–10 open of 20–22 buds or even better, C S.

433 Tropicana (Baerman) LM, good blotched commercial and garden, C.

436 W. F. Baerman (Baerman) VL, up to 12 open, wavy-petalled, C F S.

437 Germania (Salman) M, Dutch colour class winner 1972, white throat, scarlet ground, C S.

SALMON: Medium-flowered:

336 Brigitta (K & M) M, salmon-orange, C F.

333 Apollo (Fischer) M, AA, heavily ruffled, orangy-salmon blotched yellow, opens 8–10, do not overfeed, C S.

336 Headlight (Roberts) E, strong salmon, ruffled, gold picotee, 8–10 open, C F S.

335 Orbit (Butt) M, yellow blotch, formal, opens 10, C F S.

336 Scarlet Tanager (Griesbach) M, almost pure colour, ruffled, heavy substance, 'luminous' quality, 6–8 open, C F S.

337 Unique (Butt) E, darker throat, edged yellow, C F.

337 Jay-Cee (Cartmell) VL, lower petals yellow, speckled red, 10 lightly ruffled open of 22–26 buds, C F S.

SALMON: Small- and miniature-flowered:

P236 Atom (Hedgecock) M, dark enough to appear scarlet, with wide white picotee; old, essential to obtain disease-free stock, but so delightful it is worth the effort, F S.

236 Campfire (Griesbach) E, orange-toned, fine gold edging to lower petals, faint lip-markings, C F S.

236 Carnelian (Rupert) VE, wiry scarlet-salmon, 7–8 open, C F S.

237 Cliffie (Buell) M, one of the first fragrant gladioli, C F.

237 Flip (Larus) M, greenish-yellow throat, 6 open of 18–21, C F.

235 Ikon (Butt) M, deeper throat, waxy, old but still good, C F.

133 Memento (Roberts) E, face-up with creamy-chartreuse centre, 6–7 open, unique, C F.

233 Parfait (Larus) E, one of the best for form and placement, being similar to the primulinus hybrids and a good opener, S.

235 Sugar Babe (Pierce) M, blotched cream, good form, 7–8 open of 18–24 buds, C F S.

233 Starface (Vennard) M, old, but still good show and decorative, F S.

137 Small Talk (Vawter) E, old, but unique; lower petals

have wide white midribs outlined in dark 'blue', opens 6, C F.

236 Salmon Star & Red Star (Unwin), E, unique and attractive, C F.

PINK: Giant- and large-flowered:

542 Artist's Dream (Cotter) M, huge light pink, C.
445 Charisma (Baerman) M, large yellow throat, opens 7–9 in formal placement, C F S.
442 Christine (Baerman) M, well ruffled, a little short, fine for house decoration, creamy throat, 7–8 open, C F.
440 Dawn Pink (Baerman) E, pale waxy pink and white, C F.
443 Enchantress (Roberts) M, old but still good, white throat, 8–9 open, good garden flower, C F.
443 Flos Florium (K & M) M, azalea-pink blotched yellow, 8 open of 20 buds, C F S.
545 Fox Trot (Larus) M, small bright red blotch, 7–8 open, C S.
444 Friendship (Fischer) VE, still one of the best commercials, being early and consistent, its 'sports' having same virtues, C.
442 Frostee Pink (Poyner) LM, AA, delightful arrangement and florists' flower, C F.
440 Gracious Lady (Melk) M, tall, 8–9 open, cream tinted sunset rose-pink, yellow throat, slightly informal, ruffled, C F.
443 Innocent Babe (Johnson) M, 8–10 baby-pink with white centre, C F S.
443 La France (Roberts) M, AA, heavy substance, frosty glitter, white throat, opens 7–9 ruffled, formal, C F S.
546 Legend (Baerman) M, cream throat, opens 10, C S.
446 Lulu (Walker) M, top exhibition and good decorative, C F S.
543 Metropole (Salman) E, old but reliable, C S.
441 Persephone (Griesbach) LM, appleblossom-pink, white throat, tall, 8 frilled blooms open, C F S.

544 Pink Crêpe (Roberts) M, shades to white throat, big waxy, extremely ruffled stiff blooms, 7 open, C S.

442 Pink Formal (Rich) M, hint of salmon, light yellow throat, 8–10 open, C F S.

442 Pink Prospector (Baerman) LM, very good flower-form, yellow throat, opens at least 10, beautiful, C F S.

444 Pink Triumph (Labrun) E, well ruffled, healthy, good substance, C F S.

445 Powder Puff (Larus) M, tall, white throat, 8–10 open, C S.

440 Shell Pink (Roberts) M, 7–8 very pale, blending to white, C F.

446 Spic & Span (Carlson) M, an old reliable, has maintained good health, C F.

443 Spring Song (Fischer) LM, salmon-pink with yellow throat, good form, ruffled, sturdy, quite beautiful, C F S.

445 Surveyor (Roberts) M, prominent light cream throat, 8–9 open, lightly waved, tall, C S.

440 Sweet Music (Walker) LM, good pale for show or decoration, C F S.

443 Timber Topper (Melk) M, extremely tall, 24 buds, scarlet blotch, coral undertone, C S.

444 True Love (Baerman) M, 8–9 ruffled ethereal pink, C F S.

440 Vicki Lin (Bissenden) LM, 12 buffy-pink frilled blooms, best in its class, one of the greats, C F S.

447 Vista (Roberts) M, faint trace of rose, creamy-white throat, 8–9 ruffled, tall, up to 22 buds, C F S.

443 Dr Fleming (Salman) E, white throat, old, good for garden and can be brought to exhibition standard, cheap, C.

440 Pink Sweetheart (Fischer) E, good florists' flower, C.

447 Trader Horn (K & M) L, deep scarlet-pink, white feather, useful for late shows, C S.

444 Parnassus (K & M) M, 9 open, best Dutch, reliable, C F S.

PINK: Medium flowered:

343 Dresden Doll (Griesbach) E, strongly ruffled, waxy, ethereal-pink blending to pure white in throat, opens well, C F S.

B341 Melpomene (K & M) M, azalea-pink, blotched yellow, C F.

B343 Rhea Sylvia (K & M) M, deeper azalea-pink, blood-red blotch, C.

342 Village Queen (Roberts) M, tall, willowy, 8 plus open, ruffled with white in throat, C F S.

B347 Confetti (K & M) E, scarlet blotched yellow, red dusting, F.

344 Miss America (Larus) LM, AA, produces show-winning spikes all down the line, without even staking in sheltered spots, opens 12 glowing blooms; a great future if kept healthy, C F S.

PINK: Small- and miniature-flowered:

245 Camelot (Larus) E, old but still good, yellow throat, C F S.

P244 Aria, E, one of the best primulinus ever, opens 8, hard to obtain stock, F S.

243 Pinkie (Fischer) M, bright pink, white throat, rather informal, striking, C F S.

243 Pink Mini (Rich) E, 6–8 open, ruffled, good texture, C F S.

247 Sweet Debbie (Buell) E, another with a delicate fragrance, C.

RED: Giant- and large-flowered:

453 Band Wagon (Summerville) M, old, light red, only selected stock now does well, S.

458 Black Watch (Butt) M, black-red velvety, frilled, 10 open, S.

452 Christmas Red (Fischer) LM, tall, formal, lightly frilled, florists' flower, C (F?).

456 Commando (Baerman) LM, tall, 10–11 ruffled open, good substance, pales slightly towards centre, C S.

454 Courageous (Fischer) LM, 7–9 rounded blooms, prolific, C S.

455 Dixieland (Fischer) E, brilliant red, large creamy-white blotch, 7–9 ruffled open of 20 buds, vigorous, C F S.

456 Feast (Visser) M, cherry-red, grey-purple centre, cheap, C F.

456 Gail Arden (Pazderski) LM, opens 10, top show and decorative in North America for dark reds, C F S.

443 Happiness (Baerman) M, orange-red, blotched creamy-white, but opens only 5–6 on 19 buds, very decorative, C F.

458 King of Spades (Griesbach) LM, black-red, C S.

456 Olympian (Melk) M, velvety red, silver edging, C F S.

556 Oscar (K & M) LM, huge strong red, 7–8 of 20–22 buds, needs feeding, C S.

454 Dominator (Visser) M, good commercial red, best Dutch, C.

454 Red Lance (Bissenden) M, 8–9 ruffled open, C F S.

454 Red Tornado (Baerman) M, vigorous, prolific, a winner, C S.

552 Redwood (Griesbach) M, huge light red, C S.

456 Rotterdam (Preyde) M, tall plain-petalled Dutch, C S.

452 Sans Souci (K & M) LM, old but useful garden red, C.

558 Sassy Willie (Pazderski) E, top American black-red for shows and highly rated as a decorative, C F S.

454 Sequoia (Roberts) M, faint cream lines, 7–9 open of 20–22 buds, usual Roberts strong substance, C S.

554 Shirley Cole (Pazderski) M, deeper throat, 10 6in open blooms, will take some beating! C F S.

454 Sportsman (Frazee) M, 8-10 rounded velvety blooms, tall, straight, C F S.

456 Tartarian (Pazderski) M, old, but with fine record and still popular deep red, C F S.

454 Trance (Salman) M, sturdy, healthy Dutch plain-petalled, C S.

556 Winnebago Chief (Himmler) M, now old, but when grown well can be a giant; deeper in centre, S.

Page 119 (*left*) 'Sugar Babe', characteristic of the American type of ruffled small-flowered; (*right*) The type of heavy ruffling typical of many of the American and Canadian decorative gladioli that still reach show standards.

Page 120 Types of 'Peacock' hybrid. These are early-flowering and fairly hardy. (*left*) Note narrow petals and elongated blotches; (*right*) Note light spears on elongated blotches.

454 President de Gaulle (Dutch) M, 8 orange-red of 22 buds, C S.
454 Guardsman (Butt) M, brick-red, tall, 8 frilled open, C S.
455 Eclipse (Butt) M, white picotee around mid-red, 8 open, C F S.

RED: Medium-flowered:

354 Amusing (K & M) M, cherry-red, lighter in throat, C F.
353 Beacon (Butt) E, AA, salmon-scarlet with yellow throat, 8–10 open of up to 18 buds, but tall and straight, C F S.
358 Black Prince (Walker) L, velvety black-red, C S.
352 Christmas Treasure (Frazee) E, bright scarlet-red, small white line, ruffled, opens up to 9 of 20 buds, healthy foliage, C F S.
354 Frisky (Rich) L, AA, ruffled strong red with attractive white picotee, has long been a favourite, C F S.
356 Jewel Case (Rupert) E, deep red ruffled with white stamens, C F S.
358 Pickaninny (Vincent) M, 10–12 open of 21–26 black-red, C F S.
P354 Enticement (Visser), M, bright brick-red, larger prim-type, C F S.

RED: Small- or miniature-flowered:

253 Boy Scout (Fischer) LM, light scarlet with faint white picotee, informal, good performer, C F S.
252 Bugler (Roberts) VE, lightly ruffled scarlet with deeper red marks on lips, 7–9 open, formal, sturdy, C F S.
256 Cinnamon Babe (Pierce) M, orange-red outer petals, purple-red inner, 8 open, formal, C F S.
254 Corvette (Butt) E, vivid red, deeper throat, opens up to 10, old but still can be good, ruffled, C F S.
P257 Fatima (DeBe) E, blood-red with black lips and throat, F.
254 Little Slam (Larus) VE, AA, 8–9 bright red lightly ruffled open on long slender spike, C F S.
154 Dart (Roberts) E, 7 open, bright clean red, F.
252 Red Bantam (Griesbach) M, almost down to miniature size, 7–8 lightly ruffled open blooms, C F S.

253 Red Robin (Roberts) M, light picotee to light-medium red, tall, 7–8 open, C F S.

257 Tattoo (Roberts) E, dark rosy-red, striking violet-purple throat spots, 7–8 lightly waved open blooms, C F S.

257 Trooper (Roberts) E, deep red, maroon lip petals, faint gold midribs, 7–8 open, frilled and recurved, C F S.

P254 Page Boy (Fischer) M, velvety red with fine gold picotee, F.

256 Ruddiglow (Roberts) VE, nearly maroon, 7–8 open, S.

P254 Essex (Visser) E, holds 8 open in perfect primulinus placement; best prim since Aria, has been BGS Grand Champion of National Show, C F S.

P252 Frank's Perfection (Unwin) VE, the acme of the primulinus type, though not quite such a good opener as Essex, C F S.

P254 Anitra (Visser) M, red-pepper red, colour class winner in Holland, C F.

ROSE: Giant- and large-flowered:

464 Accolade (Baerman) M, good garden and commercial, ivory throat, 8 open of up to 22 buds, C.

466 American Beauty (Fischer) M, clear deep rose; opens 10, C F S.

462 Chanticler (Melk) VE, opens 8 fluted blooms, C F S.

468 Congo Song (Fischer) LM, formal silky black-red, opens about 8, S.

460 Daydream (Fischer) LM, tall, 7–8 open, ruffled, opens from tight bud, C.

466 Deep Velvet (Griesbach) M, velvety, cream spears, tall, opens 8, C S.

464 Earlibest (Melk) VE, heavy substance, ruffled, waxy, tall, white midribs, C.

462 Ecstasy (Carrier) LM, top light rose for show and decoration, 8 ruffled open of 22 buds, C F S.

464 Fond Memory (Fischer) M, salmon in throat, slightly informal, but attractive, C F.

466 His Highness (Squires) VL, sport of Director, rose-purple, 8–10 open, C S.

465 Kathleen (Melk) LM, broad silvery picotee, 9 open of 20–22 buds, C F S.

460 Lovelace (Melk) M, ruffled semi-face-up, pearly lustre, vigorous, C F S.

564 Neon Lights (Griesbach) LM, cerise, 8–10 open, but rather garish, C S.

464 Oregon Rose (Baerman) M, 7 plus open of 21 buds, beautiful colour, reliable, C F S.

463 Patricia Jean (Blanden) M, deeper picotee edge, cream throat, vigorous Australian, 8 open, C F S.

465 Peppermint Twist (Fischer) M, tall, rugged, splashes its colour around erratically, novelty value only, C F.

562 Joybells (Roberts) M, glistening 7–9 ruffled orchid-rose, C F S.

465 Persian Rose (Walker) M, lavender-rose, snow-white throat, tall, 8 open, C F S.

461 Pink Cheer (Griesbach) LM, pink picotee edge to white, 8–9 open, frilled, sturdy, C F S.

464 Radiant Rose (Fischer) LM, silver picotee, 6–8 wavy blooms open on 18–20 bud spikes, good propagator, C F.

466 Rhodora (Fischer) LM, 8–10 picoteed rounded blooms of 20 buds, vigorous and healthy, C F S.

467 Rose Tints (Fischer) LM, deep rose, dramatic blotch, C.

464 Warrior (Larus) M, fine light edge round throat petals, deeper throat, 8–10 open, C F S.

461 Deciso (K & M) M, shell-rose, darker edged, scarlet mark on yellow, C F.

ROSE: Medium-flowered:

B361 Sleeping Beauty (K & M) M, rosy-white, cream-yellow blotch, C F.

364 Curly Top (Kroon) M, short spike, darker centre, opens 8 with 6 in colour, C F.

366 Explorer (Roberts) M, deep cerise, 9–10 lightly waved, formal, C F S.

363 Fantamead (Chaffin) M, pale rose veined and dotted deeper rose, novelty value, C F.

365 Glad Hand (Roberts) VE, 7 ruffled with yellow centres, C F S.

364 Lively Rose (Park) M, lighter in throat, 8–10 waved wide-open blooms of 20–22 buds, 7 in colour, C F S.

365 Mexicali Rose (Melk) E, AA, 8–10 lightly ruffled with broad silvery-white picotee, very lovely, opens well from tight bud, C F S.

362 Orchid Rose (Roberts) M, small white spot, 8–10 ruffled, glistening blooms of heavy substance, C F S.

362 Party Pink (Melk) LM, soft rose and blue blending through pearly white, 8–10 ruffled, tall, C F S.

363 Peppermint Cane (Melk) VE, deeper rose flecks and deep rose throat, ruffled, waxy, a florist's cut-flower, C.

365 Rose Maiden (Roberts) E, 7–9 frilled, large white centre, C F.

364 Marden (Enger) LM, carmine-rose, redder in centre, 10–12 open and 6 in colour of 21 buds, ruffled, C F S.

365 Rosetone (Roberts) VE, extremely early, slightly smoky, large creamy-white centre, 7–8 waved blooms, C F.

366 Ruby (Fischer) E, tall, 7–8 saucer-shaped blooms, whippy stem, C F S.

364 Cimarron (Larus) LM, lighter in throat, feathered rose-purple, 8 open of 20–24 ruffled buds, old now, but still good, C F S.

B363 Storiette (K & M) M, rose-pink, yellow blotch, C F.

ROSE: Small- and miniature-flowered:

265 Lollipop (Baerman) VE, large white blotch, 10–12 ruffled open of 20 buds, C F S.

262 Lovely (Rich) LM, centre and most of lower petals white, 7 ruffled open of 20 buds, C F.

266 Rosy Posy (Pierce) E, tall, plain-petalled, but opens 8 or more, C S.

266 Upper Crust (Pierce) LM, old, still does well, 6 open of 17–19 buds, C.

P262 Marjolyn (Visser) M, light flame-rose, good opener, C F S.
251 Otome-No-Nozomi (Ikeda) E, soft rose with yellow picotee, C F.

LAVENDER: giant- and large-flowered (78 = purple):
475 All Aglow (Rich) L, rosy-lavender, orange centre, 8–9 ruffled open of 22 buds, tall, propagates well, C S.
572 Anniversary (Marshall) LM, low bud-count, but opens 9–10 flat round blooms with dianthus-purple coronas, C F.
477 Bess Meyerson (Walker) M, cream throat, top show and decorative, C F S.
478 Burgundy Beau (Shearer) M, rich red-purple, ruffled, 7–8 open, C F S.
473 Dawn Mist (Jack) E, blue-grey lavender, cream throat, 6–8 open, C F S.
473 Elegance (Walker) M, bluish-lavender, creamy-white throat, 9–10 open of 23–25 buds, old but still grows well from good stock, ruffled, C F S.
472 Horizon (Fischer) LM, pale rosy-lavender, AA, 10 open of 26 buds, now old, so good stock required, C F S.
578 King David (Carlson) LM, old but healthy, so cannot be left out; rich colour picoteed white, still gives show spikes, C S.
576 Lavender Masterpiece (Baerman) M, old, but still rated highly for show and decorative; good stock needed, C F S.
476 Memorandum (Salman) E, cyclamen-purple, opens 8, healthy, C S.
573 Panorama (Roberts) M, large cream centre, opens 7–8, ruffled, C S.
478 Portrait (Himmler) M, lighter in throat, 9 ruffled open of 24 buds, C F S.
578 Purple Giant (Fischer) M, 7–8 wavy-petalled of 20–24 buds, C S.
478 Purple Splendor (Baerman) L, 7–8 open, 7 in colour, 23 buds, ruffled, true self-colour, old but still does well, C S.

478 Scholarship (Klein) VL, plant early; in warm areas will give beautiful rich purple picoteed silver, 8–10 of 23 open, S.

579 Shalimar (Larus) M, rose-purple, deeper throat, opens 7–8, S.

477 Stella Dallas (Walker) M, white throat, good colour, cheap, C.

472 Antares (Visser) M, light lilac, opens 7, C F.

476 Franz Liszt (K & M) M, cyclamen-purple, narrow white line, lower petal lemon-yellow stripe, good placement, C S.

478 Britannia (Butt) M, self royal purple, 10 open, C S F.

LAVENDER: Medium-flowered (78 = purple):

B379 Brice Fair (K & M) M, cyclamen-purple, veined white, chestnut blotch on creamy-white, most unusual, C F.

375 Chantilly (Griesbach) M, ruffled and fluted, lavender mainly on upper petals, lower mainly white, 8 open of 20–22 buds, C F.

B379 Clio (K & M) M, purple with white throat, dark veined, C F.

379 Purple Moth (Pfitzer) M, darker purple throat, opens 8 of 21, usually needs no staking, C F S.

371 Crown Jewel (Fischer) E, delicate pinkish lavender-rose, ruby-red blotch, ruffled, formal, C F S.

375 Lavanesque (Baerman) M, rosy-lavender with large white blotch, good opener, reliable, C F S.

375 Lavanrose (Squire) M, sport of above, more rose, opens 8 of up to 24 buds, C F S.

375 Magic Flute (Roberts) M, light lavender-rose, large chartreuse-yellow centre, opens 7–9 ruffled with heavy substance, C F S.

379 Mark Twain (K & M) M, cyclamen-purple, creamy-yellow markings, C F.

378 Purple Plume (Roberts) M, silky pure purple, 9–10 open, lightly waved, tall, does not easily fade, C F S.

373 Troika (Butt) M, semi-face-up rose-lavender blotched

violet-lavender on creamy-white, opens up to 9, most attractive, C F S.

378 Velveteen (Roberts) VE, reddish-purple, no markings, 6–7 frilled narrow petals edged light cream, C F.

378 Wild Berry (Euer) M, deeper in throat, creamy spears, 6–8 open of 20 buds, ruffled, C F S.

LAVENDER: Small- and miniature-flowered (78 = purple):

277 Connie (DeBe) M, lilac-purple with jet-black lower petals, a 'Mini-glad', C F.

278 Grapejuice (Fischer) M, ruffled purple, especially good decorative, C F S.

277 Susie (Klein) E, old, but still highly rated blotched show and decorative, C F S.

277 Doll Dance (Roberts) E, deep lavender, large cream centre, 7–8 open, waxy, ruffled, C F S.

278 Garnette (Rich) VE, 7 open of 22 buds, medium purple, picoteed fine white, tall, informal, C F S.

176 Scout (Roberts) VE, opens 7 of 18 buds, lightly waved, reddish-purple 'tiny', F S.

278 Velvet Rosette (Rupert) M, 8 open of 21 buds, usually double-petalled, flat, well open, last well, most unusual, C F.

278 Vignette (Rupert) E, glowing purple self, 7–8 open of 20–22 buds, ruffled, C F S.

274 Camille (Larus) M, creamy-white throat, 7–9 open of 19–21 buds, plain-petalled, but still finds favour, C F S.

P279 Georgette (Visser) M, purple with lilac and orange, striped red, unusual, C F.

P278 Sussex (Visser) M, velvety purple, maroon lip-petals, C F S.

VIOLET: Giant- and large-flowered:

487 Blue Bird (Preyde) M, not to be confused with the AA 'Bluebird', a deep blue-violet, darker in the throat, opens well, C S.

581 Aquamarine (Fischer) M, light 'blue' with deep violet

throat-marks, 7–8 ruffled open of 18 buds, somewhat informal and short, but very beautiful, C F.

486 Blue Isle (Fischer) M, up to 10 deep violet frilled blooms open of 18–20 buds, very tall and vigorous, formal, C F S.

485 Blue Hawaii (Walker) LM, saturated medium violet, clean white throat, blotched violet, 9 ruffled open of 25 buds, 7 in colour, C F S.

485 Blue Night (Fischer) M, semi-formal mid-violet with deeper blotch edged white, C F.

482 Blue Ruffles (Griesbach) M, 7 or more ruffled blooms open. Ruffling is a rare feature on the larger blue-violets, C F S.

581 China Blue (Fischer) LM, AA, old now, but still among best of pale violet-blues, C F S.

484 Eternal City (Baerman) LM, a 1961 introduction that still rates among the best medium-violets for show and decoration, a tribute to its health and vigour, C F S.

482 Galilee (Fischer) LM, a lovely pastel 'blue', opens 7–8 ruffled blooms from 20 buds, formal, spike slender, plant robust, C F S.

482 Her Majesty (Turk) M, 9–11 lightly ruffled, attractive, C F S.

484 Lake Winnebago (Griesbach) M, 6–8 open on tall spikes, C F S.

485 Rippling Waters (Fischer) E, medium violet, bluer in throat, semi-formal, F S.

482 South Pacific (Griesbach) M, white throat, stock scarce, F.

483 Violet Charm (Jack) LM, really old Canadian (1953) that still does well, C F.

480 Winter Sky (Griesbach) M, 8–9 ruffled open, good decorative, C F.

580 Azurine (Rich) LM, blue-grey violet, 10 ruffled open of 28 buds, deeper flush in throat, C F S.

VIOLET: Medium-flowered:

381 Angel Eyes (Larus) M, pale violet flush with deep violet blotch, but, if reselected, original white ground-colour

can be restored for more attractive effect; consistent winner in past, C F S.

380 Blue Mist (Baerman) M, old and rather short, but beautiful and now cheap, C F.

383 Blue Pansy (Baerman) LM, large dark blotches bordered and mid-ribbed white, tall, C F.

385 Blue Velvet (Nitchman) M, deep velvety purple lip-petals, 7 plain-petalled blooms open, 22 buds, C F S.

387 Delphine (Rupert) M, very deep violet, large yellow blotches, 8 wavy open of 19–21 buds, F C S.

387 Tropic Seas (Pruitt) LM, dark blue-violet, greenish-cream lips, good opener, C F S.

386 Ultraviolet (Euer) E, very deep violet-purple, blotched deeper, opens 9 on 22-bud spike, plain-petalled, C S.

VIOLET: Small- and miniature-flowered:

283 Bluebird (Baerman) E, AA, light to medium violet, clean white throat, 8 frilled open blooms, occasionally sports lighter, C F S.

281 Linda Ruth (Griesbach) M, blue-violet, small white spear, 6–8 open of 21–25 buds, slightly ruffled, C F S.

TAN: Giant-, large- and medium-flowered:

390 Orangutan (Roberts) E, orange-tan-yellow smoky, Dracocephalus species in its ancestry; unusual, but not beautiful. Breeders might be interested.

390 Sun Tan (Roberts) E, C F S.

TAN: Small- and miniature-flowered:

290 Little Fawn (Vennard) M, 8–9 formal, deep tan, slight markings, old but still good and beautiful, F S.

290 Papoose (Griesbach) M, blend of salmon, orange, and brown, novelty, F.

290 Chipmunk (Pierce) M, variable performance, but can make good show spike, F S.

190 Table Talk (Roberts) VE, grey, rose, and tan with smoky overlay, 6–7 very ruffled open blooms, great for table decoration, C F.

290 Tapestry (Roberts) E, grey and tan with green centre, 7 heavily ruffled, very strongly substanced flowers open, tall, C F S.

290 Tokyo (Roberts) M, tan and grey, slight throat markings, 6–7 very ruffled open, medium-tall, C F S.

P291 Hastings (Unwin) E, beautiful milky coffee colour with deeper coffee-coloured blotches; outstanding, C F S.

P291 Early Hastings (Unwin) VE, similar to above, but earlier and with deep chocolate throat.

SMOKIES: Giant- and large-flowered:

494 Autumn Charm (Fischer) L, deep pink smoky with smoky red lips, rugged, F S.

598 Autumn Sensation (Grenchuck) M, huge rose-mahogany, red throat, opens 10, C F S.

497 Aztec Chief (Fischer) M, plum smoky with large white blotch, semi-formal, 8 ruffled open of only 17 buds, C F.

494 Pink Smoke (Ashley) M, rich salmon overlaid tan-gold, 8 heavily ruffled open of 20–22 buds, C F S.

497 Shady Lady (Eppig) E, blue-grey with vermilion lip-mark, and vermilion midribs, opens 7–8 heavily ruffled, F S.

594 Blue Smoke (Rich) M, now quite old, but still usually performs well, mulberry blue with smoky salmon throat, heavily ruffled, C S.

493 Gloaming (Niswonger) LM, waxy rose-smoky with smoky-yellow blotches, well ruffled, 7–8 open of 20 buds, C F S.

497 Oriental (Carrier) M, deep rose with smoky cast, large white throat, 8 open, 6 in colour, ruffled, of 21 buds, C F S.

593 Old Smoky (Fischer) E, smoky rose, white blotches, round ruffled formal blooms, C.

492 Pompeii (Fischer) E, AA, 7–8 ruffled blooms open from 19–20 buds, deep pink with lavender blend and silver overlay. When it has a good season it is really beautiful; sometimes dull, S.

497 Misty Eyes (K & M) M, brown-red-orange smoky, blotched yellow, very striking, C F.

496 Luxury (Visser) M, brown-red smoky, tinted lilac and grey, F S.

SMOKIES: Medium-flowered:

397 Colt (Baerman-Euer) M, green, flushed brownish-orange towards edges, brown-maroon throat-blotch, 6–7 open of 18, waved, F.

396 Root Beer (Griesbach) M, mahogany-brown with maroon centre (possibly should be 399), 8–9 strong-substanced winged flowers open, F S.

394 Swinger (Baerman) M, 10–14 open of 22, ruffled fire and opal, C F S.

SMOKIES: Small- and miniature-flowered:

P297 Arabella (DeBe) E, wine-red suffused brown-purple, silver picotee, C F.

296 Tampa (Visser) M, very dark grey-brown smoky, sometimes grows to 396 size, C F S.

P294 Sphinx (Visser) M, sandy brown-red veined grey, overall raspberry effect, C F S.

P296 Comic (Visser) M, browny-red, blotched deeper, strong plants, C F.

294 Blue Hue (Pierce) M, tall smoky blue-violet, fair opener, C S.

297 Davy Crockett (Roberts) M, smoky violet with large red centre, 7 of 18–19 buds open, frilled and waved, F S.

297 Gingerbread (Roberts) E, deep-toned bronze-rose, violet shadings, mahogany lip-petals edged gold, ruffled, heavy substance, 7 open on medium-tall spike, C F S.

295 Misty Morn (Roberts) E, rosy tan with smoky overcast and coral lip-petals, opens 6–7 on tallish spike, placement sometimes irregular, F.

295 Piccolo (Roberts) E, orange-bronze and rose smoky with gold picotee, opens 6–7 ruffled blooms from 16–17 buds, C F.

292 Pluto (Roberts) E, dusty rose decorative, deeper rose lip-lines, F S.

193 Rosaltha (Vawter) LM, smoky pink effect from rose over buff, deep rose blotches, an attractive 'tiny' opening 6 of 20 buds, C F S.

BROWN: Giant- and large-flowered:

599 Autumn Sensation (Grenchuck) M, rose-mahogany with bright red throat, old but rugged and still performs well in right conditions, C S.

499 Mystic Glow (Baerman) M, brown, sharp yellow throat, 7–8 open of 20 buds, wavy, C S F.

BROWN: Medium-flowered:

398 Brown Beauty (Roberts) E, light chocolate-mahogany, darkening a little at edges and in throat-lines, opens 8–9 ruffled, C F S.

398 Brown Study (Pierce) M, ruffled velvety chocolate brown, C F S.

399 Corky (Fischer) M, corky brown, flecked darker, yellow blotch, 7–8 wavy open of 16–18 buds, C F.

BROWN: Small- and miniature-flowered:

198 Chocolette (Vawter) M, solid chocolate-brown, lightly ruffled, F.

P299 Chocolate Chip (Fairchild) M, old primulinus-style with white faint picotee, good when grown well, C F.

299 Chocolate Dip (Adams) M, brown-red with smoky edging, creamy white throat, 7–8 lightly ruffled blooms open from 22 buds, C F S.

299 Little Tiger (Griesbach) E, AA, brown and orange, bronzy-red 'tiger spots', rather short and few open, F.

298 Nutmeg (Larus) E, self-descriptive, about the best in this size for show or decoration, C F S.

298 Smidgen (Pierce) M, mahogany picoteed yellow, 5–6 ruffled open of 16–18 buds, F S.

Chapter 6

Decoration and Display

About twenty years ago it was not unusual to hear the despairing cry, 'I can't do anything with these gladioli. They are too stiff and have a will of their own!' There was considerable justice in this complaint against the gladiolus at a time when both hybridists and floral artists considered its contribution to decoration as being largely restricted to massive fans of stiff-stemmed large-flowered types. The tendency to laud to the skies only those cultivars that won on the showbench by virtue of their rigid many-budded stems, with large numbers of blooms freshly open simultaneously, gave a false impression that the flower lacked versatility. Fortunately, changes have come about during the past two decades.

The changes have complemented each other. Recognition that smaller post-war houses did not call for such massive flowers as the giant-sized gladioli led hybridists to concentrate upon smaller types—such as Len Butt's 'Rufmins' (Canada), K & M's 'Butterflies' (Holland), Unwin's 'Prims' (England), Wilson's small-flowered novelties (New Zealand), DeBe's 'Miniglads' (England and Holland), and the products of such raisers as Alex Summerville in the USA.

Nevertheless, the larger-flowered gladioli still have their place in such spacious settings as churches, banqueting-halls and the foyers of large buildings. In the home, they can transform an otherwise empty fireplace into a colourful and attractive focal point, and a jug of large bright gladioli at the top of the stairs can be a breathtaking sight. Hallways, porches, patios and summerhouses may be enhanced by the judicious arrangement of no more than five or seven gladioli on their own. Many window displays in shops and stores could benefit from the complementary use of flowers, when in season.

FLOWER ARRANGEMENT

Fresh air has been blowing through the world of flower arrangement, too—not merely from the western discovery of Ikebana, but also from the development of fresh styles, such as 'expressive abstract' and 'decorative abstract'. The use of driftwood, roots, bark and other natural materials has encouraged the breeding in gladioli of unusual colours and colour combinations, especially among the smaller sizes. The flower arranger, whether concerned only with decorating the house or with entering into competition through societies and shows, now has a much wider range of colours and sizes to choose from. Particularly recommended for this work are:

G. colvillei albus (The Bride), a tiny pure white.
G. nanus Amanda Mahy, bright salmon red with small violet flakes.
G. nanus Peach Blossom, even delicate pink.
G. nanus Rose Mary, bright pink with clear scarlet blotches.

Also those marked 'F' in the list of cultivars on pp 97–132.

I should also like to see more use of the 'face-up' gladioli, especially in arrangements that will be looked at from above. These 'tiny tots'—they are usually in the 100- or 200-size groups —are becoming harder to find and may be in danger of dying out. Here are some that may still be available:

2 Cherie (Kunderd), outer petals light rose, inner dark rose.
2 Cutie (Kunderd), light rose with large deep rose throat.
2 Flower Basket (Koerner), clean pink with clear white throat.
2 Gremlin (Butt), bright rosy red, stippled in yellow throat.
1 Piccolo (Koerner), soft cream with purple shading to a red throat.

1 Red Button (Koerner), bright medium red throughout.
1 Smoky Button (Rogers), light smoky old rose with deeper rose throat.
2 The Imp (Baerman), medium yellow with sharp red blotch.
1 Thomas E. Wilson (Koerner), salmon-red with creamy throat.
1 Memento (Roberts), medium salmon, large creamy-chartreuse centre.

Preparation and accessories Spikes cut from the garden for home decoration or taking to patients in hospital can give at least ten days' pleasure. It is important to cut the spikes early in the morning, when they are fully turgid, and to place them promptly into a deep container of water. For maximum life the spikes should be cut just as the bottom bloom is opening. They may be cut 'in tight bud', ie with just colour showing but no blooms open, if a lengthy journey is involved or large numbers are to be transported together.

Many a husband-and-wife team enter shows on a complementary basis—the husband staging the straight show spikes on the bench, the wife using the not-so-straight spikes in the floral art classes. All flowers are put to good use in this way, and it is a healthy sign that more women are entering competitive spike(s) classes and more men are taking up floral art.

The novice can begin in a small way, perhaps just experimenting with home decoration at first, and later joining a club or society to pick up the finer points. Meanwhile, one can gradually add to the range of containers and accessories to broaden one's scope. The containers need to be chosen with care and some at least with a degree of originality. They must always enhance, and not compete with, the flowers. If they are to contain tall spikes, then a pinholder is essential. The old heavy lead ones are expensive, but lighter ones in cheaper materials may be bought and attached (firmly!) to the base. It is often necessary to fill the lower part of the container with small pebbles or sand to ensure stability, especially if the flower arrangement is likely to be caught in a draught.

At home, as the lower flowers die back, they can be cleaned off and the stems shortened—with a slanting cut—and smaller containers brought into play, until finally the very tips can be arranged in a shallow bowl on a dining-table or low occasional table. Thus the whole spike gives full value through its ten or more days of flowering.

The golden rule is always to create your arrangement with your own eyes at the level and in the position of the eventual viewer. Pedestal arrangements and baskets are meant to be seen in the round and at approximately head-height, though they may be organised directionally for placing in corners. Most other arrangements are going to occupy a specific position in the house or in a semi-circular or oblong recess on the showbench. Their composition must be made with this viewpoint in mind. Table bowls, which also tend to be viewed in the round, are best created by occupying a sitter's position, so that the effect may be seen as the work of art is being built up.

For show work, a collection of drapes will need to be made. Although colour is important for these background foils, so is the texture of the actual fabric. Some arrangements call for a gossamer thin, ethereal material—say, in pale blue; others require the rich plushiness of velvet—say, in purple or dark red.

The other accessories used to create the final picture, however elaborate or simple, must fit both the appearance and the mood of the total composition. Accessories can range from man-made objects that are works of art in themselves, such as porcelain figurines, to natural objects subjected to the minimum or no human interference: wood, driftwood, bark, moss, sand, fallen leaves, raffia, shells, cork—whatever you will.

Ways and means This must surely be the great delight of all flower arrangement: that you can materialise your conception in the way that you want to do it, subject only to schedule requirements if you are entering into competition. This means that you are able to take liberties with the gladiolus that would horrify an exhibitor trying to achieve the near-perfect spike. If your spikes are too long for your purpose, chop them down. If there is too much colour, nip off an open bloom or two. If you want a longer ribbon of colour the following day, pinch off the top bud to encourage faster opening. If you require just one bloom to complete a composition, select the one most suitable and chop up the spike to get it. What's a little mayhem if the result looks good?

The gladiolus is often used to give the main lines of an arrangement. This may take the form of a triangle, with all angles less than 90° or with the bottom two spikes at somewhat more than 180° to each other. Then the infilling begins. Some typical basic outlines are shown in Fig 13. However, when considering the 'lines', 'rounds', and 'flats' of an arrangement, this rather stereotyped use of the gladiolus does not necessarily have to be adhered to. The main shape can be outlined by other vertically growing subjects. Many gladioli have individual blooms so rounded—'Pink Prospector', for instance—that these can be separately used as 'rounds' in place of the more common carnations, dahlias, roses etc. Short sections cut from the stem of a gladiolus spike are ideal for wedging the whole arrangement so that nothing can topple or twist.

Use of other plants If you take up floral art seriously, you will need a ready supply of other plant material, much of which you can grow in small quantities in the garden. Many a plant worth a place there in its own right will not suffer from the occasional loss of a leaf, small branch or bloom. These are some that may be used to advantage in arrangements incorporating gladioli:

Acer (maples), particularly *A. japonicum* (ruby foliage) and its golden counterpart *A. japonicum aureum. A. Negundo variegatum*, with its variegated silvery foliage can also be effective.

Achillea (yarrow), deep yellow flowers, July–August.

Adiantum (maidenhair fern), though a greenhouse plant, is almost a must for its delicate foliage.

Alchemilla mollis (lady's mantle), often found wild in moist conditions, has lime-green flowers throughout the summer.

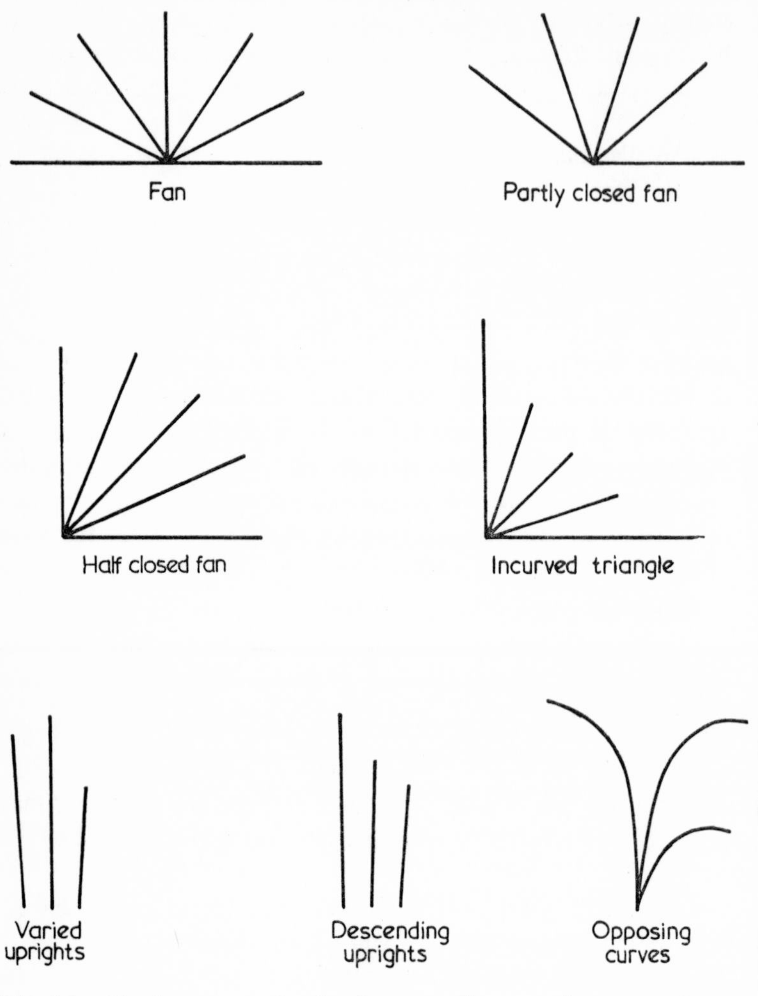

Fig 13 Simplest steps in floral art outline, moving away from symmetry

Angelica, as the foliage fades towards gold, teams in well with creams, pale yellows, and pale apricot or peach tints.

Artemesia has attractive silvery foliage that can be used to offset whites, pinks, pale lavenders, and even deep reds.

Asparagus, whose fern-like foliage needs to be used with discretion, except in traditional arrangements.

Aspidistra lurida having been removed from the front parlour window, has its leaves used much more attractively as 'flats', especially in shallow containers placed low. There is a variegated form.

Athyrium felix-femina (lady fern), an excellent alternative to the others.

Begonia, especially *B. rex*, is a common houseplant that will provide useful leaves for the base of an arrangement.

Bergenia cordifolia is invaluable for a supply of large heart-shaped leaves, some of which turn colour in late summer. Beetroot leaves, especially those from the ornamental types, could act as a substitute.

Calendula (marigold), a selection of colours to tone in with yellows and oranges.

Canna indica (Indian shot plant) needs to be raised in a greenhouse, but is often on sale at large florists. The leaves are an inspiration and should dictate the design.

Coreopsis and *Cosmea*—use like marigolds.

Cynara carduncultus (cardoon) and *C. scolymus* (globe artichoke) are both easily grown and, apart from their other uses, provide attractive leaves.

Daucus carota (carrot), the fern of which may be useful when no other is available.

Delphinium, the shorter varieties of this may be used to make the outline when gladioli are to be used as infilling.

Erigeron (fleabane), despite its horrible name, will contribute daisy-like flowers in pastel shades from pink to lavendery blue.

Eryngium, including the wild sea holly form *E. maritinum*, will supply rather spiky green and white foliage with grey-blue flowers.

Escallonia is a source of arching sprays of white or rose-tinted blossom.

Fagus sylvatica (beech) yields leaves that turn a lovely coppery colour if soaked in glycerine and these team well with all the brown and tan gladioli, as well as copper containers.

Fatsia japonica (fig leaf plant) can also be turned pale brown by soaking in glycerine and water, or may be used in their natural green or variegated colours.

Foeniculum (fennel) I grew this as a herb when I first began growing gladioli and was delighted with its feathery foliage that makes an excellent foil to almost any flower.

Francoa ramosa throws sprays of white flowers in late summer and can be useful for the later shows.

Gaillardia (blanket flower) supplies good 'fillers' of red and yellow for the more dramatic or warm-toned arrangements.

Godetia may similarly be used as 'fillers'.

Gypsophila (chalk plant), once the almost inevitable companion to sweet peas and pinks, may be experimented with as a foil to the tinier gladioli.

Hedera (ivy) All ivies and grasses have much to offer the creative exponent of floral art. A variegated ivy can even be entwined around a gladiolus spike to good effect, provided the end of a tendril with ever-diminishing leaves is chosen.

Helenium (sneezewort) is suited to the heavily dark-blotched gladioli, provided a too overall spotty effect is avoided.

Heracleum mantegazzianum (cow parsnip) Scrounge a few of the strongly serrated leaves from someone that grows it. For most gardens it is too large a plant to consider growing.

Hosta (plantain lily) These originally Asian plants will give good ground-cover and a choice of delightfully marked leaves according to which of the 20 or so species you choose.

Kalmia latifolia (American laurel) has glossy foliage with delicate clusters of pale pink frilly flowers.

Kniphofia uvaria (red-hot poker) could be employed for very dramatic purposes with tall, large-flowered yellow, orange, and red gladioli.

Ligustrum ovalifolium aureum (privet) Even the homely privet may be pressed into service if the golden-edged or variegated form is used and elegant twigs selected.

Molucella laevis (bells of Ireland) for ribbons of green shell-shaped flowers.

Nepeta (catmint) can be used as a delicate filler, provided the final effect is not too 'fussy'.

Nigella (love-in-a-mist) can add an ethereal touch to arrangements of lavenders, violets, and whites.

Onopordum acanthium (Scots thistle, cotton thistle) will 'cut and come again' for silvery-grey foil to almost any colour.

Parrotia persica (Persian ironwood) yields wonderfully glowing autumn foliage of amber, gold, and crimson for late-season arrangements.

Prunus laurocerasus (ornamental cherry laurel) will give glossy leathery leaves of mid-green to add solidity or cut out light.

Quercus (oak) Various types of oak leaves are plentiful and may well serve to cover or extend the base of an arrangement.

Rhus (sumach) In the eastern United States those species that turn the colour of their leaves in September can be teamed with the later gladioli.

Ricinus communis (the annual castor-oil plant) produces very useful large lobed and serrated leaves for really big arrangements. *R. c.* 'Sanguineus' has red-purple foliage.

Rudbeckia (coneflower, black-eyed Susan) has a deep brown-purple, almost black, centre to shades from yellow to mahogany.

Sambucus (elder) can contribute leaves of various types and colours, according to the variety: golden, yellow-green, yellow-margined, and laciniated (cut-edge).

Sansevieria (bowstring hemp) leaves can make a very effective and variegated substitute for gladiolus foliage, but they have to be coddled as houseplants.

Santolina chamaecyparissus (cotton lavender) has delicate silvery foliage that also can be used for ethereal effects.

Scabiosa atropurpurea and *caucasica* (scabious) provide infilling flowers in a range of colours.

Crambe maritima (seakale) has blue-grey leaves that make a good foil to most colours.

Senecia maritima 'Silver Dust' (ragwort) is a useful foliage plant for delicate effects.

Sorbus (rowan, mountain ash, whitebeam) is a source of twigs of various coloured and shaped leaves, often carrying berries by September.

Stachys lanata (lamb's ears or lamb's tongue) has its green leaves covered with dense silky grey hairs and can be used for special effects.

Tamarix pentandra (tamarisk) has green-grey glaucous leaves and feathery sprays of tiny rose-pink flowers in August. *T. p.* 'Rubra' has deep rose-red flowers.

Thalictrum flavum (meadow rue) has deep green leaves. The variety 'Glaucum' is the most attractive, with blue-grey foliage.

Tilia (lime tree) The various species of lime yield useful branches or twigs of a leaf or two for 'flats'.

Tradescantia is so commonly grown that the use of the variegated leaves in arrangements is often overlooked.

Typha latifolia (reed mace, bulrush) is extremely useful for 'lines' of a contrasting type to the gladiolus.

Vallota speciosa (Scarborough lily), though a greenhouse plant, can be used to team its scarlet lily-like flowers with scarlet gladioli late in the season.

Veratrum (false hellebore) has ribbed and pleated leaves, and the species *V. nigrum* produces spikes of densely packed black-purple flowers in August. NB: the rhizome is poisonous.

Vitis coignetiae (vine) is among many vines that have attractive foliage subject to colour-changing in late summer and autumn. This also bears small bunches of inedible black grapes.

Yucca (Adam's needle, Spanish bayonet) has strap-like leaves and the spikes of creamy bell-shaped flowers on *Y. filamentosa* flower in July and August.

Zea mays (maize, Indian corn) There are ornamental varieties with white or cream striped leaves; 'Quadricolor' is variegated white, pale yellow, and pink, against the green.

Zebrina pendula (wandering jew) is even more striking than variegated *Tradescantia*, the larger leaves being striped silver and backed purple.

Zinnia are half-hardy annuals in a showy range of colours that may form useful 'rounds'.

Many of the foregoing have attractive seedheads or berries before the gladiolus season has finished. These, along with fruits of all types, may be used in arrangements unless contrary to the show schedule. The final judgement must be the same as for all works of art: step back and take an impartial look at your composition. It needs to be balanced in shape and colour distribution. Perfect symmetry is the most obvious, and *dullest*, way of achieving balance; this especially applies to those four or five small arrangements on the dining-table. It should have rhythm; even if this jumps gaps, it must not be checked. As you become more expert, you will achieve successful counter-rhythms. There should be a variety of forms making one satisfying whole. There should be variety of textures that add interest and another dimension to the arrangement. All three dimensions should be adequately explored or indicated by the material involved.

If something appears to need to be added; if something appears to need to be taken out or trimmed; if something appears to need to be moved—then the work-of-art has not yet been perfected. Do the necessary! No viewer should stand in front of your creation with fingers itching to get in and fiddle!

The scope is endless. You have to start with decisions about simplicity or complexity, harmony or contrast, a natural-looking or dramatic effect. I look forward to the day when floral art does not have to confine itself to natural materials, when constructions involving gladioli in Meccano, wire, plastic building-blocks etc, will be accepted—not as a substitute for existing techniques, but as extensions to them. The creation of beauty should be the only justifiable criterion.

There was a vogue in the USA a few years ago for creating 'glamellias'—an ugly word for something quite beautiful. A corsage was made by wiring half-open, or less, gladiolus buds into the centres of fully open gladioli, thus creating a 'double' effect. Recently in Britain lapelwear has been created from a small spray of primulinus hybrid or 200- or 300-size non-primulinus gladioli: either one or two open flowers, followed by two or one buds in colour, backed by a little fern and finished with tinsel paper and a small safety pin. This has been admired wherever seen and usually leads to the inquiry, 'Are they orchids?'

It has somewhere been said that the gladiolus is the poor man's orchid. I prefer to think of the orchid as the rich man's gladiolus!

Chapter 7

History of the Hybrid Gladioli

The history of the hybrid gladioli is both fascinating and somewhat bewildering. The bewilderment is caused by two complications in trying to follow what records exist. Some of the species used in hybridising have been renamed by the botanists or were incorrectly identified initially, and some originators and purchasers of early hybrids chose to give them fancy, but botanically incorrect, names. Add to this the fact that the 'authorities' do not always agree on certain points and you may have some idea of the difficulty of sorting it all out, so I hope I have got the story straight!

To try to present matters clearly, in this book natural species that occur in the wild are given in italics (eg *G. tristis*); the results of inter-species crosses are called hybrids and treated as proper nouns with initial capital letters (eg Gandavensis Hybrids); the misuse of latinised proper names to give the impression of the discovery of a new species when actually existing species have been crossed is shown by leaving the name in roman type but enclosing it in quotation marks (eg 'G. Colvillei'); and the creation of an intergeneric hybrid is indicated by a capital X before the new name, all in italics (eg *X Gladanthera*).

In classical times Theophrastus (c 372–287 BC), the Greek philosopher and naturalist, mentions the *Gladiolus* by its Greek name (χιφιον) in *The Enquiry into Plants.* He found the plant indigenous to Asia Minor. Apparently there must have been experiments to find whether the corm had any medicinal value —I am still trying to find a safe recipe for gladiolus wine—for Dioscorides, a Greek physician of the first or second century AD, listed in his *Materia Medica* a gladiolus, the description of which fits *G. communis*, still found around the Mediterranean.

The present name appears to have come from Pliny the Elder (AD 23–79), a Roman officer and statesman turned scientist, who used it in his *Naturalis Historia*, the only extant work from his voluminous writings. This is the earliest record of its Latin name, which is usually attributed to the shape of the leaves. However, this form of leaf is common to many plants of the Iridaceae family, so I consider it much more likely that the emerging flowerhead reminded the Romans of the short sword (*gladius*) used by their soldiers, which similarly gave us the word 'gladiator'. Pliny had been a soldier, and *gladiolus* is literally 'little sword'—and very apt, too, if you look at the shape of the tip of a Roman sword.

Attempts were also made by apothecaries in Western Europe in the sixteenth century to use the plant medicinally—with what results we do not know. Two species were grown in gardens for ornamental purposes, but fell into disfavour as more attractive flowers became available.

Fresh interest was created in the genus *Gladiolus* when the Dutch East India Company established its victualling station at Cape Town and began a garden there. To stock it, seeds, roots, corms and bulbs of any attractive plants were collected from the interior. Some viable plant material was sent to Holland, along with paintings of the plants and their flowers. In the days before photography, these were hand-copied, enabling Dr Leonard Plukenet, of Westminster, in England, to publish figures and descriptions of *G. alatus* and another South African gladiolus as early as 1692.

Doubtless ships calling at the Cape brought some seed or corms to England direct. By the eighteenth century, South African species were growing in English gardens along with the commoner European *G. communis* and *G. italicus* (syn *G. segetum*). Linnaeus saw *G. angustus* in the garden of George Clifford of Hartecamp, enabling him to publish its first description in 1737. By his death in 1778, Linnaeus had also described *G. alatus*, *G. recurvus*, *G. tristis* and *G. undulatus*.

Philip Miller received seeds from a Dutch friend, Dr Job

Baster, from which he grew *G. tristis* and other South African species in the Chelsea Physic Garden from 1745 onwards. *G. cardinalis,* one of the earliest contributors to the modern race of gladioli, was described and figured by William Curtis of the London Botanical Garden and also in Holland virtually simultaneously.

Hybridisation began early in the nineteenth century. The Hon and Rev (later Dean) William Herbert is known to have been making crosses in 1806; but his hybrids, including the most famous, 'G. pudibundus', were apparently all sterile, being probably triploid in character. The firm of James Colville and Son of King's Road, Chelsea, had more success. A cross between *G. cardinalis* and *G. tristis* var. *concolor* produced 'G. Colvillei'. This is an example of a personal name being latinised and used as though it were the name of a species. Robert Sweet, the foreman of Colville's nursery, described and drew this hybrid in 1823. It later threw a sport with white flowers and coloured anthers. This in turn sported to the all-white 'The Bride' about 1871. 'G. Colvillei' corms and cut-flowers are still sold today. The corms are planted in the autumn for spring flowering. The 'roseus' and 'ruber' forms preserve something of the *G. cardinalis* colour and markings.

The House of Schneevoogt, in Holland, crossed *G. cardinalis* with *G. blandus* (now renamed *G. carneus*) to create a hybrid again misleadingly named as though it were a natural species —'G. ramosus'. A fellow Dutchman, Ernst H. Krelage, developed a race of these, better termed Ramosus Hybrids. The backcrossing of this hybrid with *G. cardinalis* produced dwarf early-flowering gladioli. It is believed that this was the lineage of the race known as 'G. nanus', better referred to as Nanus Hybrids. These have been developed in Holland and the Channel Islands, and cultivars called 'Spitfire', 'Peach Blossom', 'Nymph', and 'Blushing Bride' are on sale as cut-flowers in the spring.

The first summer-flowering hybrids came from a cross between *G. psittacinus* (as it was then termed, now regarded as one of the variable forms of *G. natalensis*), a large-flowered yellow flaked orange, and *G. cardinalis*, a smaller red with three narrow long white blotches on the lower three petals. However, here the authorities differ. Philip O. Buch claims that all the hybrid progeny were red, so that the second parent must have been *G. cardinalis*; but he gives no source for checking his statement. G. Joyce Lewis and A. Amelia Obermeyer state that the illustrations of the early Gandavensis Hybrids show no trace of *G. cardinalis* influence (presumably including no sign of the characteristic elongated blotches seen in the Nanus Hybrids; but the whereabouts of these illustrations is not stated) and conclude that *G. oppositiflorus*, a tall-growing white striped amethyst from Natal, was used.

If Lewis and Obermeyer are right in their supposition, this would help to explain the many reversals of flower form that occur in modern gladioli, as the other species involved are all double-lipped. Some of the opposite-facing tendencies of *G. oppositiflorus* would have been bred out, but the forward-looking double-row appearance may have been achieved only by some individual blooms virtually turning upside-down.

Their collaborator in *Gladiolus: A Revision of the South African Species*, Dr T. T. Barnard, supports their assertion, stating that *G. oppositiflorus* is believed to have been sent to England by John Forbes early in the nineteenth century and also that it had been grown in Europe from 1823, though under the erroneous name of *G. floribundus*. Certainly *G. oppositiflorus* would have contributed added height and an increase in bud-count and the number of fresh flowers open at one time.

Whatever the truth is, this was the crucial cross in the history of summer-flowering gladioli. It was made by the Belgian Hermann Josef Bedinghaus, head gardener to the Duc d'Aremberg, and he later started his own nursery in Belgium. The late-flowering character of *G. natalensis* gave summer-flowering progeny that were named Gandavensis Hybrids. Most of the stock was sold to Louis van Houtte, who introduced it. Soon many other nurserymen were not only listing these progenitors of our modern gladioli, but were also intercrossing them to create new cultivars. Eugene Souchet, at Fontainebleau, was able to raise many good ones by open pollination and some of these were displayed in quantity in Paris when Queen Victoria and Prince Albert visited the Emperor Napoleon III and his Empress in 1853. The Queen was so impressed that she gave instructions for these gladioli to be grown at Osborne House. (Their distant descendants were handed in welcoming bouquets to the first returning Russian cosmonauts.)

William Hooker had meanwhile made a successful inter-species cross in England—*G. blandus* (now *G. carneus*) × *G. psittacinus* (now *G. natalensis*)—and called the results Brenchley-ensis Hybrids, after his home town. John Standish, the nursery-man at Bagshot who had supplied the corms to Osborne House, not only produced Gandavensis Hybrids equal to the French ones, but also crossed the Brenchleyensis Hybrids with *G. cruentus*, a blood-red parent that introduced amaryllis features into its progeny.

In France, Victor Lemoine of Nancy crossed *G. purpureo-auratus* (now reclassified as a form of *G. papilio*) with Ganda-vensis Hybrids and introduced some 152 named Lemoine Hybrids, again incorrectly referred to as 'G. Lemoinei'. He had been a pupil of van Houtte at Ghent, after which the Ganda-vensis Hybrids had been named. From his Lemoine Hybrids he developed a new line named 'G. Nanceianus' after *his* home town, by crossing with *G. cruentus*. William Bull of King's Road, Chelsea, had been advertising corms of the South African species *G. papilio*. Lemoine now used these for obtaining crosses with his Lemoine Hybrids to produce the so-called 'blue' gladioli, of which the best known was 'Baron Joseph Hulot', present in the ancestry of some of our modern gladioli, especially the purples, violets, and lavenders.

In Germany, Max Leichtlin, aware of Lemoine's work, tried the other cross, *G. cruentus* on Gandavensis Hybrids, in 1877 and started a line first termed Leichtlini Hybrids. These he sold, undeveloped, to a French firm that in turn sold them to John Lewis Childs, of Floral Park, New York. He named them 'G. Childsii', but we will call them Childs Hybrids. Although the early Ramosus, Gandavensis, and Lemoine Hybrids had all previously been introduced into the United States, the American

Civil War had retarded hybridising there. Childs went to work in his Long Island nursery and from 1893 introduced some 150 named cultivars that were the genuine Childs Hybrids.

A fellow-American, Dr W. W. van Fleet, raised 'Gladiolus Princeps' from *G. cruentus* and a Childs cultivar; but probably the most outstanding story is that of A. E. Kunderd. Amos E. Kunderd, born in 1866, produced the first new strain or race for some twenty years when he introduced his 'Kunderdii Glory', the first with ruffled flowers, in 1903. This paved his way to commercial success and his continued experimental breeding added the laciniated type in 1923, another twenty years later. Seven hybridisers each paid $1,000 a corm when it was first released—and that was in dollars with the buying-power of the 1920s!

The Kunderd Gladiolus Farm at Goshen employed 100 people, grossed $325,000 a year from corm sales alone, and exported to forty-four countries. For twenty years the business flourished and Mr Kunderd became the first president of The American Gladiolus Society. The financial crash of 1929 ruined his firm, like so many others; but he struggled through the years of the Depression, refusing to have himself declared bankrupt, as he believed it would be unfair to his creditors. He lost all that he had worked so hard to build, but even in his later years he was still developing miniature gladioli. The British Gladiolus Society had decided to have a special gold medal struck in honour of Mr Kunderd's 100th birthday. Acting on advice from America, they brought forward their plans a year, but fate still forestalled them—A. E. Kunderd died shortly before his 99th birthday. Nevertheless, his work lives on in the traits of some of today's gladioli.

Yet another breeder-extraordinary was the late Ralph Baerman. The number of his introductions is legion—one has only to look at any NAGC lists giving the raiser—and his flowers were notable for their beauty. This time the BGS did manage to bestow its highest honours before it was too late: an honorary life membership and the BGS gold medal for services to the gladiolus. Ralph was the only man I ever heard of who managed to make a living simply by breeding gladioli. His expert knowledge was sought to hybridise flowers that would perform satisfactorily in winter in Florida and also stand trucking to all parts of the United States and even Canada.

One cannot leave the recent development of the gladiolus in the USA without mention of at least two more names among the many deserving recognition. Carl Fischer's hybridising success may be judged from the fact that the All-American patented gladioli, so thoroughly tested on such varied trial grounds, included thirteen Fischer originations among the first twenty-four released. No other breeder has come near that record. Secondly, Winston Roberts—because he has concentrated more upon strong texture and good ruffling in his originations—has probably been underrated because many of his gladioli are a little short in the bud-count to be show-winners. However, for sheer beauty, weather-resistance and suitability as cut-flowers, his numerous introductions take some beating.

Reverting to Britain, James Kelway united his derivatives of Souchet's hybrids with Lemoine's Nanceianus Hybrids. The distinctive race was termed 'G. Kelwayi', in other words Kelway Hybrids, later ones of which were called 'Langprims' after the firm's base at Langport in Somerset. These latter were a rather intermediate type between the wide-open large-flowered and the smaller hooded primulinus hybrids.

The term 'primulinus hybrids' is almost certainly a misnomer, but undoubtedly is with us now to stay. *G. natalensis* has three yellow forms, one of which is *G. primulinus*, collected and identified by Baker. However, something was imported into England about 1902 that was given the name *primulinus*. This must have been *G. nebulicola* from the vicinity of the Victoria Falls, as maintained by Capt Collingwood Ingram, who identified this form of *G. natalensis*, since this is the only one of the three forms said to have been established in cultivation. Whether this is the same as what was sent to Kew for trial around 1890 is not certain, but it probably was, for the Kew trials report was quite prophetic: 'It ought to be the starting point of a new race of garden gladiolus.'

Although many were quick to incorporate the lighter features of this new discovery into their breeding programmes (Kunderd, Kelway, Lemoine, Vilmorin, Pfitzer, Thorburn, Cayeux, Le Clerc, Unwin etc), the recognised primulinus hybrid type was nearly lost to cultivation during World War II, at any rate in Europe. It took the devoted efforts of Frank W. Unwin, of the firm of W. J. Unwin Ltd, to re-establish a large range of true-breeding 'prims' in Britain. From this he more recently developed his Star range, with long narrow widely held petals and opening buds that form three-pointed stars with their tips above the open blooms. In 'Frank's Perfection' Mr Unwin achieved what he regarded as 95 per cent of his concept of a perfect 'prim'. Other true primulinus hybrids have come from P. Visser Cz at Sint-Pancras in Holland, notably 'Essex', which made history by being the first of its type to be declared Champion Spike of the Show at a BGS National Show.

The other species gladioli that have been incorporated into the genetic 'pool' of the garden gladioli, though not to so far-reaching an extent, are: *G. dracocephalus*, used by Saunders of Reigate, Surrey, on the Gandavensis Hybrids in 1889 (its dusky maroon and green have helped to extend the colour range, but this is another that has now been revised to being a sub-species of *G. natalensis*); *G. saundersii*, a reddish-orange blotched white, with ground colour stippling on the blotches, that added extra size, though this is somewhat similar to, and has been confused with, *G. cruentus* and thus may have been what Leichtlin used in his crosses; *G. aurantiacus*, golden-yellow or orange and yellow, partly stippled or streaked with red, used by Lemoine with Leichtlin Hybrids; and *G. cooperi*, a yellowish ecotype (variant found in a specific area) of *G. natalensis*.

In Germany, Wilhelm Pfitzer at Stuttgart, also a pupil of van Houtte, engaged in the early breeding involving the first hybrids and his firm produced improvements down the years. Unfortunately, all professional breeding of gladioli in Germany has recently ceased.

Not all advances in gladiolus types have come through the use of species. Occasionally a helpful sport occurs or an unexpected new form shows up among seedlings from a cross between cultivars. Leonard Butt, of Ontario, Canada, had just such a piece of good fortune at the end of World War II. A 'Brightsides' × 'Roi Albert' cross threw up an interesting orange-red seedling that was small in flower-size and had even ruffling all round each bloom. This he named 'Crinklette' and by breeding from it he obtained a whole race of what became familiarly termed 'Rufmins', though they are more strictly small-flowered non-primulinus gladioli.

The trend towards smaller sizes was brought to Holland by the breeding work of Arie Hoek, on behalf of Konijnenburg en Mark N.V. Using 'Ladykiller' as both seed and pollen parent with small-flowered gladioli from America and Canada, and later using extensively 'Tarantella' and 'Biedermeier', a race with pronounced contrasting blotches was produced that became aptly named in trade and popular parlance 'Butterfly' gladioli. These were nearly all of 300 size and have become common throughout Europe and are even sought after in America.

Recently attempts have been made to produce earlier-flowering gladioli to bridge the gap between the spring-flowering and the summer-flowering types. Frank Unwin has used crosses between his own primulinus hybrids and the Nanus Hybrids to produce one seedling in the F_3 generation (the 'great-grand-children') that had unmistakable Nanus traits. This was then crossed with his best primulinus hybrids and his newer Star range to give a race named 'Peacock' gladioli. These are not only earlier-flowering, they have also inherited increased hardiness. Selected 'Peacocks' that I have left *in situ* for three seasons (ie overwintered twice without protection) have multiplied and have had to be dug up for better spacing. Almost all have *G. cardinalis*-shaped blotches, but in a wide variety of colour contrasts. They may also prove to be more disease-resistant than other summer-flowering types.

A different version of this breeding has been done by the Dutch firm of Konijnenburg en Mark, using Nanus and Colvillei Hybrids on their well-established 'Butterfly' range. The seedlings introduced to commerce are called 'Coronado' gladioli and again most feature blotches, particularly purple ones. They will flower in late May to June in a good season in southern Britain, or during June in a cool one, and so tend to be even earlier than most of the 'Peacocks'.

Another approach has been to attempt to breed from plants of different genera. This can only be successful if the chromosome-count is identical or an exact multiple, though freak results might be obtainable with differing numbers. In the 1930s Collingwood Ingram crossed *G. tristis* var. *concolor* with *Homoglossum watsonianum* and produced a series of hybrids that have proved to be completely hardy in Kent and Gloucestershire. He named these *X Homoglad* and his best selection he called 'General Smuts'. Philip O. Buch took a further step by crossing one of these hybrids on to the *G. natalensis* sub-species *G. psittacinus*; but only Dr T. T. Barnard, by inter-breeding and selecting Homoglad Hybrids, has produced scented seedlings this way.

The attempt to produce scent in summer-flowering gladioli has been a long and continuing one. For decades the hybridists' dream was to achieve a viable cross between *Gladiolus* and *Acidanthera*, which is perfumed. This proved difficult, because summer-flowering gladioli are tetraploid (60 chromosomes), whereas *Acidanthera* is diploid (30 chromosomes). However, the possibility was there, as each had a basic number of 15. The likelihood was, however, that any resultant plants from such a cross would be sterile triploids (45 chromosomes).

In 1955, Mrs Joan Wright of North Auckland, New Zealand, already a breeder of horses, successfully crossed seven *Gladiolus* cultivars with pollen from *Acidanthera bicolor* 'Murielae'. From the resultant seed, plants grew from six of these crosses. (The full story may be read in *The Gladiolus Annual, 1966*.) Over 220 plants had flowered without a trace of scent and the project was about to be abandoned when a *Gladiolus* 'Filigree' × *Acidanthera* seedling came into flower; not only was it fragrant, it had the *Acidanthera* characteristics somewhat modified towards *Gladiolus* traits. This white seedling proved to be a triploid; by further crossing, a tetraploid—'Lucky Star'—was raised that was considered a genuine bigeneric and christened *X Gladanthera.* This delicately scented newcomer will cross readily with ordinary gladioli.

There are several genera closely related to *Gladiolus* and various bigeneric hybrids have been created, but those from *Acidanthera* and *Homoglossum* are the only ones that have found a place in horticulture. Because they can be made to cross with each other and for various more technical reasons, plants in the related genera have been classified as *Gladiolus* from time to time. With the current intensive study of these plants by South African botanists such as Obermeyer, Goldblatt, and Marais, much of the previous confusion has been eliminated. The genus *Acidanthera* has now been formally disbanded by these and other specialist workers. Consequently, the plants previously known as *Acidanthera bicolor*, *Acidanthera bicolor* 'Murielae', *A. murielae*, and *A. tubergenii* 'Zwanenburg' become *Gladiolus callianthus*, *G. callianthus* 'Murielae', *G. callianthus* 'Murielae', and *G. callianthus* 'Zwanenburg' respectively.

While this new classification means that *X Gladanthera* will need to be eventually renamed, it in no way diminishes the achievement of Joan Wright in successfully transferring fragrance from a diploid species into the tetraploid garden gladioli.

Homoglossum is also under scrutiny by botanists and may go the way of *Acidanthera*; but, for the moment, the *X Homoglad* hybrids are regarded as being bigeneric.

Mention must be made here of two other major efforts aimed at creating scented gladiolus hybrids. For some twenty years Dr T. T. Barnard hybridised at Wareham in Dorset the diploid South African species from Cape Province. Nearly all the crosses have involved the scented species *G. alatus*, *G. carinatus*, *G. caryophyllaceus* (*G. hirsutus*), *G. gracilis*, *G. liliaceus* (*G. grandis*), *G. tristis*, and *G. virescens*. He has produced a very large colour range and polyploidy (triploids and tetraploids) has arisen spontaneously. These scented hybrids have been variously called *G. tristis* hybrids and Purbeck Hybrids. All require cool greenhouse treatment for the best results (ie a minimum temperature of 7° C), but most will survive and flower for a year or two in sheltered sunny spots in southern England. The best diploid cultivars include 'Christabel', named after the late Lady Aberconway, wife of the president of The Royal Horticultural Society, while the best tetraploids include 'Corfe Castle' (*G. grandis* × *G. caryophyllaceus*) × (*G. caryophyllaceus* × *G. grandis*). These and many others have received Awards of Merit from the RHS and Haarlem.

However, the scented gladioli that have been the most successful commercially are those raised by the Rev Dr Clifford D. Buell and the late Rev Spencer. Selected only from the existing garden gladioli available at the time in the USA, several have a distinct rose perfume unlike any scent found in the wild species. The best known of these cultivars are 'Acacia', 'Thisisit', and 'Yellow Rose'.

Interest in breeding for perfume in gladioli has been revived with Joan Wright's breakthrough and hybridists are intercrossing the three main fragrant races mentioned above.

For the possibilities of future breeding, see Chapter 9.

Chapter 8

Hybridising

For anyone wishing to try his hand at creating new flowers, the gladiolus is the ideal subject. The reproductive organs of the flower are easily recognised and easily accessible; the act of making a cross (fertilising one cultivar with pollen from another) is simple; almost all gladioli are lavish with the amount of pollen they create; the sequential opening of the flowers means that cultivars can be crossed that are up to a fortnight out-of-phase in their flowering times, and the later opening of flowers from small stock extends the period of pollen-availability; large numbers of seeds are obtainable from any one plant; gladiolus seed usually germinates readily without any special conditions having to be artificially created; the existing cultivars have a complex ancestry that gives variety among the progeny; and there are still over a hundred wild species that have not been used to contribute characteristics to the cultivated hybrids. Add to all this the fact that you begin seeing the results two years after crossing, when the better grown seedlings will come into flower, and it becomes obvious that this is a field in which anyone can experiment without feeling that it will take a lifetime's work to 'get anywhere'.

TECHNIQUE

To make a cross, choose two parent cultivars with desirable characteristics that you hope to combine. One will act as the seed-parent and can be regarded as the female in the union; the other will supply the pollen and is thus the male. Of course, insects will beat you to it if you let them, but you can usually beat the insects without complicating matters too much.

When the male parent is coming into flower, cut the spike and place it in water indoors. This is to stop pollen from being lost at the crucial time as a result of rain washing it away or wind causing much of it to drop off. Bagging of the female flower often causes rotting owing to the bag becoming rainsoaked or condensing transpired moisture. A simpler way to keep insects out is to fasten the petals together with a safety-pin or put a cigarette packet over the bud before it opens. By the time the flower should have been open two days, the three-fronded stigma (see Fig 1, p 15) will have become sticky and receptive. However, there is some danger that the bloom may be self-pollinated, as the pollen on the anthers also becomes ripe about this time. Two courses can be followed. Often the stigma is sticky during the first day the bloom would be open and therefore the pollen from the chosen male can be applied then. Should the stigma not be ready on the first day, the bloom can be emasculated by breaking off the three stamens by their filaments and discarding the anthers before they can release pollen, and then reclosing the flower with the safety-pin or the cigarette packet.

No brush is needed for transferring the pollen. If a fine paint-brush were to be used, as is sometimes advocated, then after one cross it would have to be dipped into pure alcohol to kill any remaining pollen before the next lot was loaded on it and then the brush would have to be washed to rid it of the toxic alcohol. Nature's own materials will not be injurious to each other. Go to the male spike and find three anthers with ripe pollen just beginning or about to fall on to the petals below. Using a pair of flat-ended tweezers, grip each filament in turn and carefully snap it off, shedding as little pollen as possible. Carry the filament between thumb and forefinger to the seed-parent. Uncover the enclosed stigma and wipe the 'hairy' surface of the parted fronds with the anther, pollen side down. Smother all the length of each frond with pollen grains. If there is not sufficient on one anther, use a second and even a third. Provided enough ripe pollen coats the stigma, there need be no fear that later insect-borne pollen will fertilise the ova.

Try to pollinate the lowest flowers on a spike and do not

expect to grow more than four to six seed-pods to maturity, even though a plant will often do so if it remains green until late in the season.

Label the seed-parent plant that has been crossed on to, so that it does not inadvertently get cut before running to seed. Small white card price-tags, already threaded with strong cotton, are ideal for this, as the names of the parents can be written on one side and the numbers of the blooms pollinated (counting from the bottom) on the other side. The same details should be entered into the garden notebook at the start and brought up-to-date daily, as follows:

Date	*Name of seed-parent and number in row*	*Name of pollen-parent*	*Blooms crossed*
1/8/75	AURORA 6	× GOLDEN PHEASANT	1, 2, 3, 4

Once sufficient blooms have been crossed to ensure the quantity of seed you desire, the stem should be snapped off about six buds higher than the last bloom crossed. All the blooms and buds not deliberately pollinated should be nipped off clean at the stem. These two procedures will encourage the plant to run to seed early, prevent its energies going into chance seed from later insect-pollination of higher flowers, but allow for any die-back of the mutilated stem to stop well above the seed-pods retained.

The plant should continue to be treated as a growing entity and in the autumn each green seed-pod, on reaching maturity, will turn brown and start to split down three of its seams. A careful watch needs to be kept for this as, if the remaining stem is not cut in time, the seed could be scattered by the wind before you are aware of the danger. As soon as any one of the pods shows signs of splitting, cut the stem just above the foliage and carry it indoors. Tick the notebook entry to show that this cross has been harvested.

If the right moment has been chosen, these stems will not need to stand in water. Retain the seed-pods on them as long as possible, allowing the sap to dry out as the stems stand in short vases, mugs, or tankards, keeping dry air circulating around the pods. As soon as there is any danger of seed falling, carefully break off the pods into any suitable receptacle (cereal bowl, cigar box, tobacco tin etc) and place the price-tag label with them. They may now remain anywhere dry and moderately warm until a convenient time for packeting.

Small manilla envelopes are suitable for storing seed and recording on the outside the year and the details of the parentage. Copy these from each label and place the envelope under its relevant receptacle. Once you are sure the seed is thoroughly dry, the pods can be pulled apart into three segments and the seeds rubbed loose with a thumb. Only those seed-wings with a small brown ball inside contain a viable seed; any that are quite flat or have a black centre should be discarded along with the debris from the pod. A pinch of fungicide may be dropped inside the envelope as an added precaution before sealing. Thereafter the envelopes can overwinter in any dry but not too warm place, such as a drawer in the bedroom, a desk or a bureau, or inside a box in the corm-store.

Gladiolus seed may be sown quite early in the next year, from February onwards, if a warmed greenhouse or at least bottom heat is available to encourage early germination. The seeds are quite big compared to most flower seeds, so that a common fault is to sow them too shallowly in insufficiently deep trays. For the maximum growth during the first season, boxes at least 6in (150mm) deep are required. The bottom 4in (100mm) should be filled with good friable soil enriched with well-rotted compost and a sprinkle of bonemeal. Above this a layer of seeding compost 1in (25mm) deep is spread and then the whole box is thoroughly soaked. The seeds are then pressed slightly into the damp surface at 1in intervals in each direction. Immediately the sowing is completed, a ½in (13mm) layer of very fine river-sand (sharp sand) is poured over the whole area. I find this the best way to keep seed from being washed to the surface during subsequent waterings or exposure to rain and at the same time it gives minimal resistance to the first seed-leaf. Indeed, such fine sand can be tamped flat to ensure that every seed is in intimate contact with the compost and yet the surface drainage will still be good. When the seedling foliage is well developed, it is possible to add yet more fine sand to give the tiny corm additional depth, which encourages the growth of the main corm rather than the production of cormlets.

Each box should be labelled with the parentage or a code number that reveals this on reference to the garden notebook. If two small batches of seed are sown in one large box, a clear division should be made between them by using a strip of thin wood or plastic, and the front and back of the adjacent rows of seed from different crosses should be marked by plastic labels bearing the cross number or details for each, to save confusion when harvesting. If fortnightly waterings of weak liquid feed have been given throughout the period of growth and the boxes have been placed where they will receive maximum sunlight, some of the resultant corms will be as much as 20–25mm in diameter; most of the corms will flower during the next season, with at least sufficient blooms to indicate whether a seedling is worth retaining and growing to a full-sized spike before seriously evaluating it. The harvested corms are treated in the same way as for named cultivars, except that there is no point in retaining any cormlets until one is certain that the new seedling is worth propagating or that one wishes to ensure its continuity in order to use it for further crossing.

THEORY

Gladiolus hybridising can be considered as falling into three categories: (1) dabbling, (2) semi-scientific breeding, (3) scientific investigation.

For the ordinary gardener a little dabbling is worth a try, can be good fun, does not require the keeping of elaborate records, and may—such is the complex ancestry of the garden gladiolus—result in a show-winner, a new line, or a weird freak. You can cross any gladiolus with any other that takes your fancy or just happens to be out at the same time, or even let the insects do the choosing and gather seed from open-pollinated blooms allowed to set pods, sow it, persevere for three years until all the seedlings have flowered, and see what you get. At least they will all be your gladioli of your own raising and, if you finally decide to throw them all away, little will have been lost. On the other hand, there is a fair chance of seeing one or two goodish ones that you would like to keep and propagate, even if only to be able to say that they are of your own raising. The chances of originating a world-beater first time are very slight; but if enough people dabble, the odds that someone will succeed are considerably shortened.

At the other extreme is the sort of hybridising that can only be satisfactorily done as a research project in a horticultural institute or a university department of botany. Whatever its object, such scientific investigation often takes many years to complete, may involve a whole team of specialists, and requires the continuous keeping of meticulous notes, as well as the ability to interpret the results correctly. Such investigations may yield valuable information about colour-inheritance, whether disease-resistance is linked to other characteristics, whether and by what means perfume may be strengthened in the few summer-flowering gladioli that are mildly fragrant etc. However, it is beyond the scope of the part-time amateur to follow a programme that leads from a few initial crosses to the crossing of some hundreds of seedlings among themselves or back on to the original parents, with the recording and analysis of all the data provided each season. Done thoroughly, this would quickly escalate from thousands to millions of seedlings with which the amateur has neither space nor time to cope. By the use of existing knowledge of genetics, statistical techniques and a computer, possibly much of such work could be condensed into fewer actual crosses for each generation and still yield valid results—but this is still the realm of the full-time scientist with adequate resources.

SEMI-SCIENTIFIC BREEDING

Most serious hybridising of the gladiolus seems to me to fall into this second category. This is neither the casual 'hit-or-miss' crossing of anything that happens to be in bloom when the hybridist walks among his gladioli nor can it be quite as thoroughly scientific as the purist among scientists would like. Nevertheless, it does try to build upon what science already offers, modified by the predilections and hunches of the actual hybridist. It is a combination of art and science: art because the hybridist has to have an eye for beauty, the imagination to visualise the end-products he is seeking, and the foresight to judge what the public will find attractive; science because, unless he works systematically towards his ends, the hybridist is unlikely to achieve them and is merely a dabbler on a larger scale.

Anyone wishing to make significant progress in gladiolus breeding within a decade or two needs to master two areas of knowledge: genetics and ancestry.

Genetics All living things contain genes which, astonishingly for their tiny size, have a complex code within the molecule of the chemical Deoxyribonucleic Acid, usually abbreviated to DNA. This molecule is a double-stranded helix—that is, it looks like two spirals overlapping one above the other, as though two extended springs had been intermeshed. The number and sequence of its constituent parts determines whether a cell is directed to become part of an eye or part of a fingernail, part of a root or part of a petal. We have no direct control over individual genes; they can only be unpredictably affected by radiation or other chemicals, usually with disastrous results, as in the case of the thalidomide babies.

The genes are attached in thread-like groups termed chromosomes, because these bodies (somes) can be stained (*chroma* is Greek for 'colour') and seen through a powerful microscope. The chromosomes can to some extent be controlled during breeding, so that a rather crude control is established over which genes are passed on.

The smallest number of chromosomes that a gladiolus tissue cell can have is 30. These are derived from two sex-cells each containing the basic number (15), indicated by botanists by the small letter 'n'. Therefore the simplest form of gladiolus is a diploid (2n = 30) and many species are of this type. At meiosis (the splitting of a cell into halves or gametes), the pollen grains are formed with only one set of chromosomes (15); likewise the ovules (eggs) have only one set. The two sets combine on fertilisation to give each seed its full 30 and thereafter the multiplication of cells by mitosis (division) is accompanied by a longitudinal splitting of each chromosome so that the full genetic code from 30 chromosomes enters each new cell before the dividing wall is completed. Thus each cell is fully programmed so that the plant will grow into its individual self and no other.

However, the modern summer-flowering garden gladioli are all tetraploids—that is, they have four times the basic number of chromosomes in each cell (4n = 60). They therefore inherit two sets of chromosomes from each parent (2n or 30). This complicates matters, but also leads to greater possibilities of variety and therefore of improvement or fresh characteristics. The important point to remember is that the gene or combination of genes responsible for any factor (earliness, height, number of buds, cormlet production, colour, hardiness, or whatever) will be present not once or twice in the seed-cells, but four times.

This is where some understanding of Mendel's laws of inheritance is essential. Although Gregor Mendel knew nothing of genes and chromosomes when he conducted his experiments with garden peas, by observation and statistical analysis he was able to explain the mechanics of heredity, and his laws, with slight modification, still hold good today. They enable us to recognise the effects of the genes though we cannot see the genes themselves, to calculate the odds against achieving certain desirable objects, and thus to plan a breeding programme that avoids the necessity of crossing everything with everything and ending up with more seed than can possibly be grown on to flowering. The theory is best illustrated in a grid diagram. If the four genes or gene groups for a given factor were all different, which is the most unpromising start one can have, then the plant could be represented as ABCD for this factor. Self-pollinating it would be a cross denoted as ABCD × ABCD. As two of the chromosomes bearing these genes come from the pollen and two from the ovule, all possible combinations can be pictured thus:

From seed-parent:	AB	AC	AD	BC	BD	CD
From pollen-parent:						
AB	AABB	AABC	AABD	ABBC	ABBD	ABCD
AC	AABC	AACC	AACD	ABCC	ABCD	ACCD
AD	AABD	AACD	AADD	ABCD	ABDD	ACDD
BC	ABBC	ABCC	ABCD	BBCC	BBCD	BCCD
BD	ABBD	ABCD	ABDD	BBCD	BBDD	BCDD
CD	ABCD	ACCD	ACDD	BCCD	BCDD	CCDD

This is the F_1 generation, ie the children of the original parents. Six out of every 36 seeds are as genetically 'mixed up' (ABCD) as the original parents and on flowering will be similar to them for the given characteristic. All the other 30 have had at least one gene 'dose' doubled. Let us assume that A is the factor it is desired to strengthen. (The same argument and method would apply to any other letter.) Nine out of 36 (a quarter of all seeds) have a double dose. Even if this were recessive, so that in AABB and AACC and AADD the other double doses were dominant and 'masked' the effect of AA, nevertheless six of these nine, on flowering, would exhibit strengthened A characteristics.

Now cross any pair of these six (or all the lot, if you have the facilities) or self any one (or again all). One example will suffice to show what happens, say AABC × AABC. All possible combinations are:

From seed-parent:	AA	AB	AB	AC	AC	BC
From pollen-parent:						
AA	AAAA	AAAB	AAAB	AAAC	AAAC	AABC
AB	AAAB	AABB	AABB	AABC	AABC	ABBC
AB	AAAB	AABB	AABB	AABC	AABC	ABBC
AC	AAAC	AABC	AABC	AACC	AACC	ABCC
AC	AAAC	AABC	AABC	AACC	AACC	ABCC
BC	AABC	ABBC	ABBC	ABCC	ABCC	BBCC

This is the F_2 generation, ie the grandchildren of the original parents. Note that one seed in 36 has inherited purely A characteristics (AAAA) and eight others have three As in their constitution, so that if A were recessive it would still show up clearly in nine out of every 36 spikes; if it were dominant, the incidence could be 27 out of every 36. Moreover, the single AAAA in each 36 would be true-breeding; if selfed, all its progeny would contain AAAA. Thus there could be an F_3 generation reliable for this factor.

The above procedure is known as 'in-breeding' and superficially appears to achieve the desired results quite quickly. However, the selection and crossing were for one factor only. The danger is that, while one characteristic is being strengthened, other undesirable characteristics may likewise be being encouraged. This crossing of sister-seedlings is often termed 'sibbing', but as with humans and animals too much inbreeding often has damaging effects. The inbreeding of gladioli is usually accompanied by loss of vigour, a proneness to disease, and eventual early death of most of the seedlings.

However, we started from the most pessimistic premise that the original parents would have only one 'dose' of genes for the desirable characteristic and that this might even be recessive. Usually the hybridiser starts with plants that already display to a marked degree the characteristics he wants in his seedlings. Then, if the original parents are robust, the worst effects of sibbing may be ameliorated by crossing the best of the F_1 generation on to the original parents. This is termed 'back-crossing' and may be incorporated into 'line-breeding', in which the original parents and the best of several generations of their progeny may be involved in a complex programme of sibbing and back-crossing.

'Out-crossing' is the commonest form of hybridising for new cultivars and simply means that unrelated or very distantly related parents are used, mainly to ensure hybrid vigour in the seedlings. It is a good idea to use parents raised in different countries to achieve hybrid vigour coupled with an extension of the variation of genes, thus making distinctively new types and colours possible. Canadian originations can be crossed with Australian, Russian with American, Dutch with New Zealand etc.

Ancestry This is where ancestry, the second body of knowledge, becomes important since, with the relatively easy worldwide availability of the best gladioli, it may be that the parents or grandparents have something in common after all. The North American Gladiolus Council registers and produces long lists of parentages of gladioli that have been introduced commercially. All raisers, in whatever country, should supply the NAGC archives with full details of any gladiolus marketed.

For an ancestry as complex as that of 'Jade Ruffles', a linear chart becomes too cumbersome. Even a circular chart does not entirely solve the difficulty of the geometrical progression of a 'family tree': 2 parents, 4 grandparents, 8, 16, 32, 64 etc, but on a large sheet of card or stiff paper the full details can be shown. Fig 14 goes back nine generations from 'Jade Ruffles' to 'Richard Diener' × 'Salbach's Prim' (see pp 168–9).

The information yielded by such a chart can be considerable, especially if one becomes knowledgeable about the older cultivars. This one shows, among other things, that—in the creation of 'Jade Ruffles'—'Picardy' entered at least seven times directly and three times as a parent of 'Wings of Song', and the remote ancestor 'Beacon' was also a 'Picardy' seedling with pollen from 'Mrs T. E. Langford', so that it contributed at least eleven times to our limited knowledge. To anyone not already aware of the fact, this would indicate that 'Picardy' is a good parent. Indeed, it is famed for its progeny. In the late 1950s I started to compile a list of registered cultivars with 'Picardy' as one of the immediate parents. Working alphabetically, before reaching B, I had fifty-eight names beginning with A. At that point common sense prevailed over enthusiasm and I abandoned the task. However, note that 'Picardy' was used not in early crosses alone, but only four generations back in the 'Doll House'-'Yellowstone' line. Although it was first introduced by Professor Palmer in 1931, 'Picardy' is still commercially available and its potential for hybridisers is by no means exhausted.

Another interesting fact is that 'Elizabeth the Queen' turns up as both a grandparent and great-grandparent of 'Preview'. On the seed-parent side it is only three generations away from *G. nebulicola* × a Childs Hybrid. Outstanding in its day, 'Elizabeth

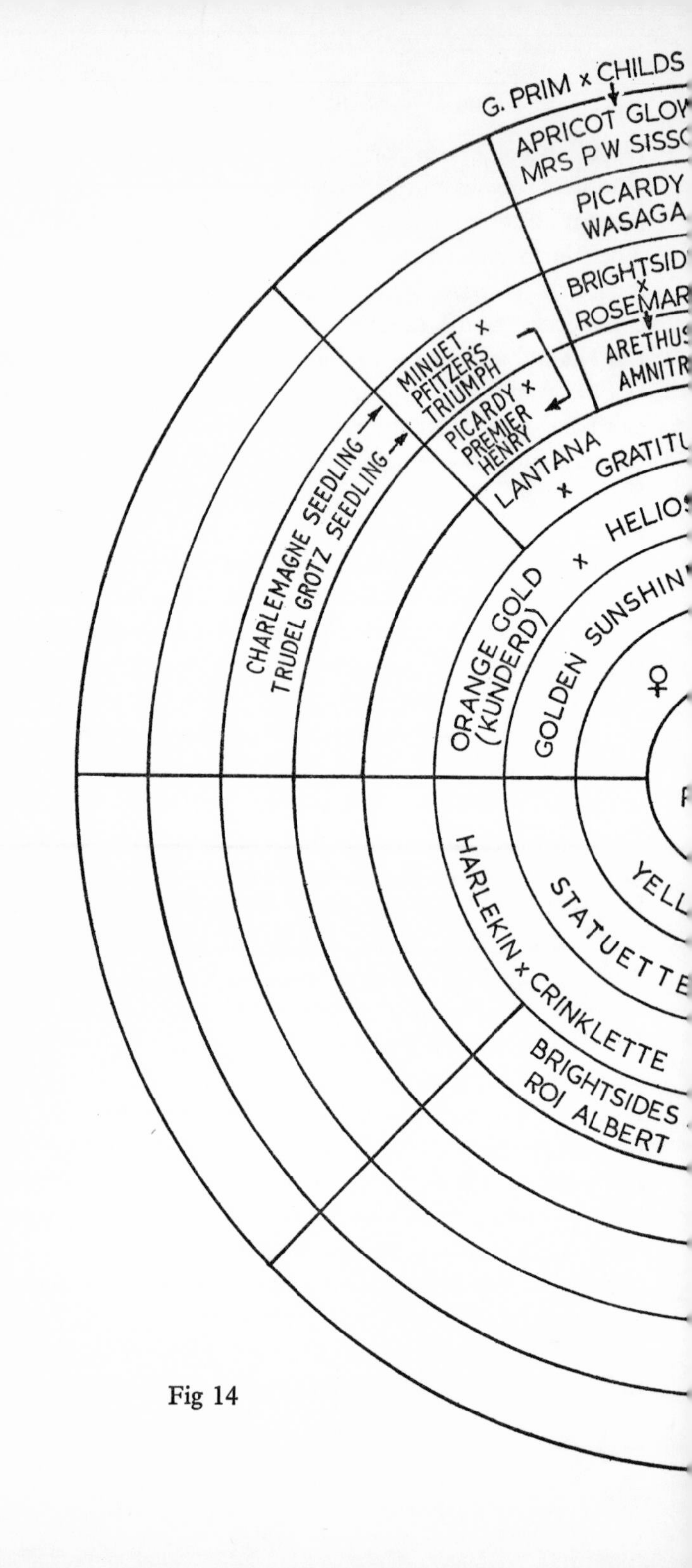
G. PRIM ×
PICARDY
WASAGA
MINUET ×
PFITZER'S TRIUMPH
PICARDY ×
PREMIER
HENRY
LANTANA
×
CHARLEMAGNE SEEDLING
TRUDEL GROTZ SEEDLING
×
ORANGE GOLD
(KUNDERD)
GOLDEN
HARLEKIN × CRINKLETTE
STATUETTE
BRIGHTSIDES
ROI ALBERT

Fig 14

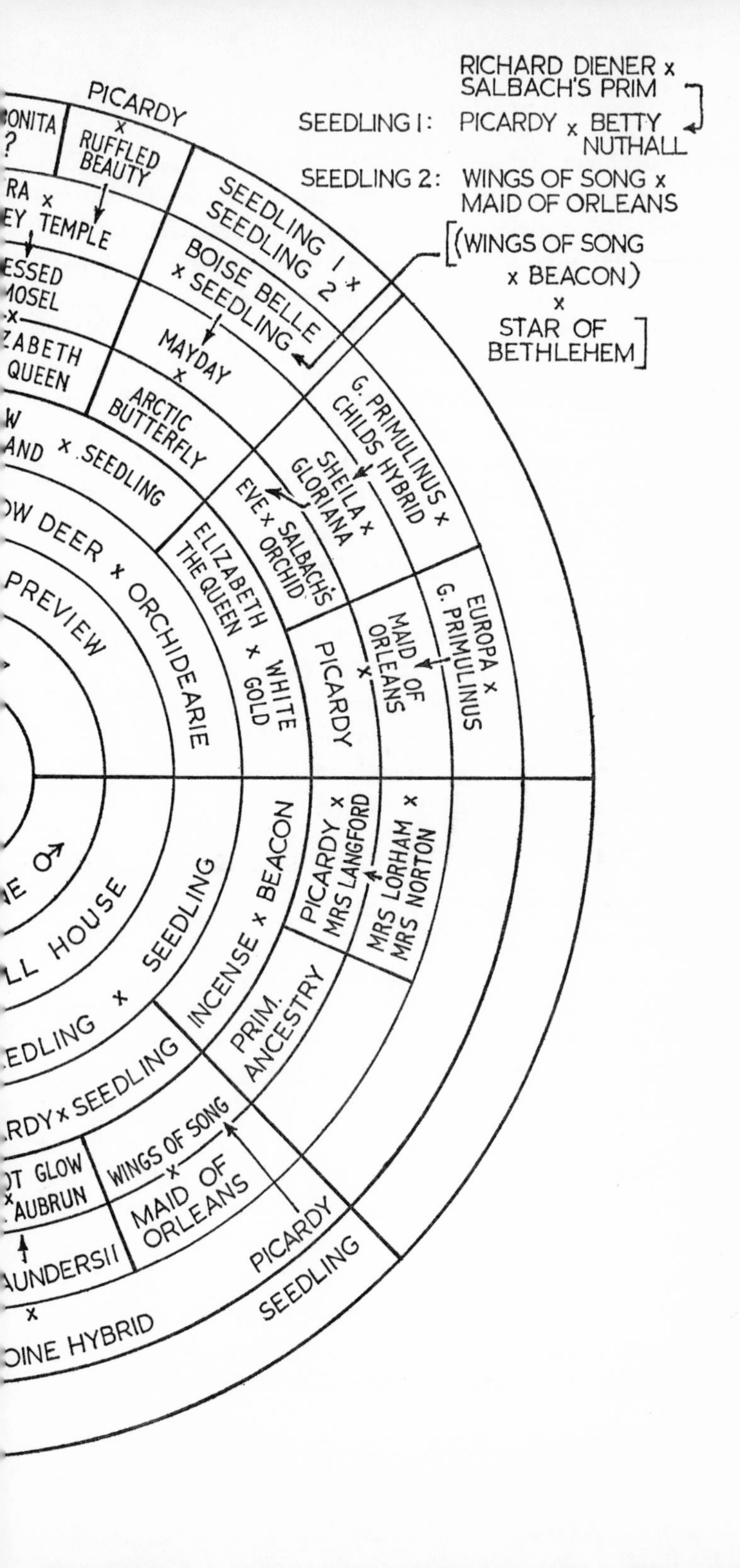

RICHARD DIENER x SALBACH'S PRIM
SEEDLING I: PICARDY x BETTY NUTHALL
SEEDLING 2: WINGS OF SONG x MAID OF ORLEANS
[(WINGS OF SONG x BEACON) x STAR OF BETHLEHEM]
PICARDY x RUFFLED BEAUTY
TEMPLE
SEEDLING I x SEEDLING 2
BOISE BELLE x SEEDLING
MAYDAY x
ARCTIC BUTTERFLY
x SEEDLING
G. PRIMULINUS x CHILDS HYBRID
SHEILA x GLORIANA
EVE x SALBACH'S ORCHID
ELIZABETH THE QUEEN x WHITE GOLD
x ORCHIDEARIE
PREVIEW
EUROPA x G. PRIMULINUS
MAID OF ORLEANS x
PICARDY
PICARDY x MRS LANGFORD
MRS LORHAM x MRS NORTON
INCENSE x BEACON
SEEDLING
x
PRIM. ANCESTRY
x SEEDLING
WINGS OF SONG x MAID OF ORLEANS
PICARDY
SEEDLING
x

the Queen' shows that it is not a long journey from using species material to producing show spikes. Moreover, the Primulinus (*G. nebulicola*) contribution can be seen to enter from several directions. To this is added an enormous variety of genetic possibilities, since this incomplete ancestry chart shows that in the 'family background' are *G. Saundersii*; Lemoine and Nancy Hybrids developed in France (Charlemagne, Roi Albert, Emile Aubrun); a Gandavensis Hybrid developed in Belgium (Europa); Childs Hybrids and Kunderd named cultivars (Ruffled Beauty, Orange Gold) developed in the USA; two German originations (Pfitzer's Triumph and Rosemary Pfitzer); and the first 'Rufmin' from Butt in Canada (Crinklette). This should ensure not only vigour and good health, but great variety in any further seedlings from 'Jade Ruffles'.

Only very rapid use of outstanding cultivars could give such a complicated ancestry in such a relatively short time (the 1931 'Picardy' was used eight generations back), so most cultivars' charts would be much simpler to construct. This is true of another famous parent, 'Innocence', which via 'Friendship' has 'Maid of Orleans' and 'Picardy' as great-grandparents.

Procedure What is the best procedure for a modern hybridist, possibly a novice to breeding? First he needs to clarify in his own mind the major characteristics that he intends to breed for. He then needs to select from the existing gladiolus cultivars at least two displaying these characteristics strongly, or one showing them strongly and another with them less pronounced, but coupled with other highly desirable characteristics.

A semi-scientific programme needs to be followed for several years. Suppose his two selected plants are Y and Z; SP means used as seed-parent and PP means used as pollen-parent. Then his first step would be to make the cross both ways: Y(SP) × Z(PP) and Z(SP) × Y(PP). The resultant seed should be kept separate.

Theoretically, there are equal chances of growing good progeny whichever way round the cross is made; in practice, some cultivars prove better at setting seed and make better seed-parents than others, which tend then to be reserved as pollen-parents.

Once the F_1 generation come into flower, the possible number of crosses is multiplied enormously. At this stage as many as possible should be made, to encourage recessive characteristics to appear—hence the common sense of starting with two cultivars only. From two to three years later the F_2 generation will be in flower and things will seem to have become frightfully complicated. The choices of further crosses now are:

$Y \times F_1$; $Z \times F_1$; $Y \times F_2$; $Z \times F_2$; $F_1 \times F_1$; $F_1 \times F_2$; $F_2 \times F_2$.

To go as far as the F_3 generation is desirable, as many unusual things turn up at this stage, including quite new 'breaks' that may be 'throwbacks' to species ancestry. However, the whole thing might be imagined to have got quite out of hand by now, with tens of thousands of seedlings to cross among one another. Fortunately, there are two safeguards against this.

Nature has its own safeguard in that 'inbreeding' tends to increase any proneness to disease. Well before the F_3 stage, many seedlings may have died off or earmarked themselves for destruction for lack of vigour in growth or poor keeping qualities in store. Consequently the total viable stock will have reduced itself naturally.

The hybridist will also be adding to this selection process. After the F_1 wholesale crossing, he will hybridise only among those plants that strongly exhibit his own desired objectives coupled with good vigour and the absence of any pronounced defects. He may produce his 'winners' from sibbing or back-crossing. By careful record-keeping, what he will have learnt will be the full potential of his original parents; the increase or decrease in disease-resistance that back-crossing gives, as opposed to sibbing (thus revealing the original relative healthiness of his two selected parents); and which of his F_1 and F_2 seedlings are worth incorporating into a breeding programme because they are highly reliable in passing on certain strengthened characteristics. Often such an unnamed seedling will be utilised as the parent of a whole series of good commercial or show cultivars.

If, apart from our original choices Y and Z, another pair having in common a different desired characteristic had proceeded simultaneously through the same programme, there would now be two 'lines' to blend to achieve outstanding results. However, every additional line doubles the amount of time, work and space needed, transferring the exercise from the 'semi-scientific', within the scope of the amateur, to the realm of the full-time professional scientist.

Trials Anyone having produced new gladioli that they think are worthy of wider appreciation and possibly commercial introduction has two possible ways of bringing his/her products to the attention of the interested public (apart from the expensive business of advertising). One way is to exhibit the cultivar against the best in commerce at major shows. This was done in 1973 by Eric Anderton of Bury, Lancashire, at the British Gladiolus Society's Annual Show at Newcastle upon Tyne. His spike was taken from the seedling classes to the court of honour and there adjudged the 'Champion Spike of the Show'. Eric's current ambition is to 'sweep the board' at Southport Show with exhibits consisting solely of seedlings of his own raising.

As it is difficult to have spikes at their exhibition peak on show dates, it is better to have seedlings tested in trials at test gardens. This is particularly true for those newcomers not intended as top exhibition cultivars, but meant to be judged as commercial cut-flowers, for garden decoration, or for use in floral art. Moreover, anything with claims to fragrance needs to be sniffed at various times of the day—early morning, in the heat of the afternoon and late in the evening.

After making advance inquiries about requirements and conditions, stock for trial may be sent to:

THE TRIALS SECRETARY, THE BRITISH GLADIOLUS SOCIETY, PASHLEY FARM, NINFIELD ROAD, BEXHILL-ON-SEA, SUSSEX, ENGLAND.

The British Gladiolus Society tests stocks from all over the world in a minimum of three test gardens spread across Great Britain and gives three awards—in ascending order: Certificate of Commendation (CC), Award of Merit (AM), and First Class Certificate (FCC). Raisers are encouraged to submit for a second trial any cultivar gaining one of the lower awards. The First Class Certificate can be awarded only to an entry that has previously been granted an Award of Merit and has reached that standard a second time.

THE TRIALS SECRETARY, ROYAL HORTICULTURAL SOCIETY, WISLEY GARDENS, SURREY. Here one test garden only is used—the famous RHS Wisley Gardens. The awards are similar to those of The British Gladiolus Society, except that the lowest award is 'Highly Commended' (HC).

EXECUTIVE SECRETARY, ALL-AMERICA GLADIOLUS SELECTIONS, 5402 VILLAGE WAY COURT, AUSTIN, TEXAS 78745, USA. The AA trials take place in some twenty-eight test gardens, mostly in the USA, including one in Hawaii; and a few in Canada. From two to four new introductions are selected each year and are patented by AAGS, so that once they are released they are sold under licence at a minimum charge for each corm until the patent expires.

The Royal Netherlands Gladiolus Society conducts annual colour class competitions, but these are for Dutch commercial raisers only.

Chapter 9

The Future for Gladioli

The future of the gladiolus lies partly in the hands of the hybridists and partly in those of the gardening public. The breeders may try to foresee what might appeal to the public, particularly if they are hybridising for a commercial outlet. Some, however, pursue their own particular developments, often in the hope that these will catch on, if only with a limited range of specialist aficionados.

The public at large—even the gardening public—take little interest in the remarkable variety of available gladioli that this book has tried to bring to their notice. They are too easily satisfied with what the florist offers—often a parody of a first-class gladiolus, usually due to forcing—and the limited choice offered by garden centres, chain stores and general horticultural catalogues of some of the oldest cultivars extant that have been propagated by the million. There needs to be a compromise between big business money-making on the one hand and the preservation and development of a wide choice of beauty on the other—and only you, the buying public, can bring this about. Ask for, press for, better cultivars and a larger selection. What you cannot obtain through traders, seek from national, regional and local societies, and even by individual correspondence. If your 'ideal' does not seem available anywhere, set about breeding it for yourself!

With such a wealth of beauty now produced for us, it may seem churlish to point out that mistakes have been made in the past and may still continue unless trends alter. The very rapidity with which more beautiful cultivars and new colours have been produced, particularly in the USA, has been too often accompanied by a neglect of that fundamental feature: good health.

Hundreds, probably thousands, of originations have proved disease-prone within half-a-dozen years and disappeared ignominiously. This has tended to give the gladiolus a bad name as an unreliable genus. Even stock that with careful cultivation could have remained healthy has acquired and spread disease as a result of the uncaring attitude of some mass-propagators that treat plants merely as a way to make money and take only a short-term view.

Fortunately, there is now a widespread and growing awareness of the dangers inherent in this attitude. Selected Glads Inc, in the USA, give their All-America award to from two to four new cultivars yearly that have passed stringent tests in over twenty trial gardens, and allow the patented cultivars to be propagated only under licence. Even with these precautions, some of the earlier AA originations varied in health according to the supplier.

In Britain it is likewise possible to take out a copyright on a new gladiolus under the Plant Varieties and Seeds Act, 1964, and receive royalties for cut-flowers and corms sold by licensees. Unfortunately, the Act came into operation just at the time when the commercial breeding of the gladiolus in Britain was declining almost into non-existence, and the costs of going through the thorough testing needed to acquire such a copyright are beyond the amateur's pocket, unless he has guaranteed outlets that allow him to recoup his expenses from commercial royalties.

In Holland, many firms have long concentrated upon producing large-flowered self-coloured or nearly self-coloured cultivars for distribution across Europe. The actual bloom did not matter so much as the ability to produce masses of clear colour or spikes suitable for standing singly in German front-windows and looking impressive for several days. To their credit, the Dutch have tried to ensure good health and in many cases preserved it, although stock imported from overseas has brought with it problems of preventing the spread of infection. The Dutch were late to appreciate the extra dimension and added beauty that ruffling gives to the gladiolus, and so have predisposed their markets and the European public to acceptance of plain-petalled flowers as the norm.

The emphasis on ruffling in North America has in some cases led to the production of 'exotics' that are heavily pleated and curiously spurred, often to the detriment of the flower-form. Whilst there should not and indeed cannot be any arbitrary limitation to experiments, it is to be hoped that the criterion of beauty, as opposed to novelty and weirdness, will be applied in deciding what to retain and develop.

It is ironic that such new forms should be pursued in North America when an old distinct form with a very respectable pedigree is not recognised there by any authoritative bodies. I refer to the primulinus hybrids that are so useful in the garden because they have whippy stems that rarely break in normal winds and need no staking once the bases have been heeled in or hilled up to ensure that the plants are growing upright. This type, well epitomised by 'Aria' and 'Nadia' of old, has a grace of its own. Some North American originations are of this type, but the failure to recognise it as a special category has led to varying degrees of mongrels between it and the ruffled miniature- and small-flowered, whether of the informal Butt type or more formal type.

A further feature that seems to have been largely ignored is that many of the ancestors of modern garden gladioli were quite hardy and could be over-wintered in the ground wherever it did not freeze to the depth at which the corm lay. Now the virtue of hardiness is being resought, though I venture to suggest that it is present in many more currently available gladioli than most growers suspect. Provided an envelope of sharp sand is present to prevent the corm from rotting, many gladioli need not be lifted until thinning out is required. This not only reduces the chores of lifting, drying, cleaning and storing—not to mention replanting—but restores the gladiolus to the life-cycle of the original species.

Perhaps we should think ourselves lucky that the gladiolus is as tolerant and co-operative as it currently is. There should be small surprise that not all of the much-vaunted cultivars live up to one's expectations, especially in the first year of planting and flowering. Think for a moment what we do. Apart from the unnatural business of storage and transporting, we are usually removing a plant from one habitat to another, involving changes in soil, climate and daylight hours. Ideally, plants should be left in the habitats in which they were reared and selected because of their outstanding performances. To judge gladioli fairly, since few of us breed and select our own for local conditions, we should at least grow our own stock from cormlets and flower these before discarding any as unsuitable to our environment.

For gardens that suffer from heavy summer rains, often whipped along by the wind, I would suggest that a further range of gladioli would prove suitable, along the lines of my own seedling 'Spokesman'. These would be 'short-handled', opening their flowers just above the foliage to prevent stems from snapping; small- to medium-flowered, to avoid too much wind-resistance; and heavily ruffled with strong substance to the petals, to avoid rain damage. For many, this would be the ideal garden flower and probably not need staking. When the bottom blooms have 'gone over' and the top ones are starting to open, the stripping of the flopping flowers would provide enough stem for cutting for indoor decoration. These would be a 'dwarf-formal' type.

To give a better cutting crop, some breeders are now trying to ensure that three, four or more spikes grow from each corm planted. The new 'Coronado' (K & M) and 'Peacock' (Unwin) lines are achieving this by the use of Nanus and Colvillei hybrids in the parentage, but with a slight loss of bud-count. Rubbing out the apical bud on planting encourages more of the side-buds to develop into spikes.

Whether summer-flowering gladioli should have fragrance bred into them is a much-debated question. Some argue that the gladiolus is satisfactory as it is in bloom when many quite strongly perfumed flowers—eg roses and carnations—are available. However, it is noticeable at shows how many of the less-knowledgeable public, women in particular, sniff at large-flowered exhibition gladioli to find out whether they have any scent. Much painstaking work has been done to produce perfume in summer-flowering gladioli, on the grounds that this would add yet a further ingredient to their attractiveness. What has so far been achieved is included here for the benefit of those wishing to pursue this line of breeding or merely to enjoy the work of others—though it must be pointed out that some people cannot detect the subtler fragrances that others swear they smell!

The work of Mrs Joan Wright in producing the mildly fragrant *X Gladanthera* has been alluded to in Chapter 7. Whether this type of fragrance can be combined to strengthen that from other sources is yet to be adequately proved. There is always the possibility that the alleles contributing the *Acidanthera*-type fragrance may in some way be mutually incompatible with alleles contributing other types of fragrance. Additionally, Dr J. E. Amoore believes that man can smell seven primary odours (F. A. Hampton thought there were ten) and that each has its own characteristic receptor sites in the olfactory membrane in the nose. Different types of fragrance might 'blend' without increasing the apparent strength of scent. Nevertheless, some form of fragrance is claimed to have been detected in the following gladioli (summer-flowering):

335 Gay Flame (Zeller) M, salmon, blotched scarlet on cream in yellow throat, 7 open of 24 buds.
337 Gay Heart (Zeller) M, deep salmon, yellow blotch, red arrows, 7 ruffled open of 20 buds.
333 Highland Miss (Zeller) E, light salmon pink upper half, creamy yellow lower with scarlet arrows on lip-petals, 9–12 open of 22.
441 Acacia (Buell), large-flowered creamy-pink, red spot.
432 Bouquet (Buell) M, ruffled salmon, cream throat.
237 Cliffie (Buell) M, coral-salmon, red spot.
333 Gaytime (Zeller) M, salmon, red spot.
464 Gwen (Pickell), large-flowered rose.
413 Jimmy Boy (Buell) M, light yellow, red throat-marks, 7–8 lightly ruffled open of 15–16 buds.
301 Perfume (Braver), medium-flowered white with rose dart.

465 Spice (Buell) M, medium lavender, cream throat with lavender markings, 8–9 ruffled open of 20 buds.
420 Happy Talk (Zeller) L, pale orange, heavily ruffled, opens 11.
Sweetie (Spencer), red with white throat.
Pink Fragrance (Spencer), medium-flowered pink.
247 Sweet Debbie (Buell) E, ruffled dark pink, cream throat.
Yellow Rose (Spencer), large-flowered yellow with red throat-mark.
400 Angel's Breath (Buell) LM, cream-white, 7–8 wavy open of 17–19.

Many people are now trying to combine the virtues of various mildly fragrant gladioli without propagating undesirable characteristics. There are some in the 'blue' range coming along that can trace their fragrance—in part, at least—back to Dr McLean's 'Viola Sweet' (GE–13). Others are coming from the *X Gladanthera* 'Lucky Star', though it is proving difficult to lose the *Acidanthera* form, but new colours are appearing. Whether you are breeding for fragrance or not, it is a good idea to take a deep sniff at every new gladiolus that opens—first in the morning, then in afternoon sun, finally near sunset. It would be a pity if a potential contributor to the 'fragrance pool' got away undetected!

However, where breeding is concerned, I suggest that certain priorities should be observed: 1 Health, 2 Hardiness, 3 Vigour, 4 Beauty, 5 Ease of propagation, 6 Weather-resistance. The other characteristics that make a good commercial gladiolus are secondary to these.

For the production and maintenance of healthy gladioli we really ought to consider moving back to organic methods of horticulture, and also reducing and finally eliminating the use of toxic sprays and dusts. A healthy plant will shake off a mild attack from disease or insects; a sickly plant should be uprooted and burnt, not coddled. Good compost-making would cure most of the ills our gardens are prone to, and the use of 'old-fashioned' natural insecticides and fungicides is probably the only answer to the manner in which insects and fungi develop resistant strains to every new man-made 'miracle-worker' put on the market and making its contribution to world pollution.

Finally, what sort of a future world are we wanting, anyway—one in which everything becomes standardised until there is little choice, little variety, left? Well, whatever happens in other spheres, you could always find choice and variety among gladioli if you wished. It is up to you.

APPENDIX A

List of Useful Addresses

BRITAIN

Beneglads, 30 Sunningdale Road, Flixton, Urmston, Manchester M31 1DT

Cramphorn Limited, Cuton Mill, Chelmsford, Essex CM2 6PD

DeBe Miniglads, 65 Gaston Way, Shepperton, Middlesex

Kelway & Son Ltd, Langport, Somerset

W. J. Unwin Ltd, Histon, Cambridge CB4 4LE

Walter Blom & Son Ltd (of Holland), Leavesden, Watford, Herts

Plant Variety Rights Office, Murray House, Vandon Street, London SW1

Hon Sec, British Gladiolus Society, 10 Sandbach Road, Thurlwood, Rode Heath, Stoke-on-Trent, Staffordshire.

Hon Treasurer, British Gladiolus Society, 31 Rupert Avenue, High Wycombe, Bucks

Hon Sec, Scottish Gladiolus Society (BGS), 63 Gardiner Road, Edinburgh EH4 3RL

Hon Sec, Gladiolus Breeders' Association, 13 Chelsea Avenue, Thorpe Bay, Southend-on-Sea, Essex

Hon BGS Trials Sec, Pashley Farm, Ninfield Road, Bexhill-on-Sea, Sussex

C. Sales & Son, 226 Chingford Mount Road, Chingford, London E4 (for exhibition vases)

HOLLAND

Secretary, Nederlandse Gladiolus Vereniging, Divonalaan 43, Hillegom

NV Konijnenburg en Mark, Offenweg 42, Noordwijk-Binnen. (Wholesale only)

A. J. Preijde Gladiolus Specialist, Breezand
NV H. J. Salman & Zonen, Lijnbaanweg 37, Noordwijk
P. Visser Cz., Benedenweg 212, Sint-Pancras
Van Tubergen, 86 Koninginneweg, Haarlem

EAST GERMANY

VEB Deutscher Landwirtschaftsverlag, 104 Berlin, Reinhartstrasse 14

LATVIA

Aldonis Verins, 10 Valgales Street, Riga 17, Latvian SSR

USSR

A. N. Gromov, NIIOH, Perlovskaya Station, Mitishchi 19, Moscow Region

JAPAN

Takayoshi Asakawa, 8 Gakuen-Higashicho, Kodaira, Tokyo

SOUTH AFRICA

National Botanic Gardens of South Africa, Kirstenbosch, Newlands, Cape Province

AUSTRALIA

South Australian Gladiolus Society (Mr Ellis, 60 Mitcham Avenue, Lower Mitcham, South Australia 506Z)

NEW ZEALAND

Mrs Joan Wright, Box 395, Dargaville

WEST GERMANY

Wilhelm Pfitzer, 7012 Fellbach/Wurtt, Postfach 37
Deutsche Dahlien- und Gladiolen-Gesellschaft, 674 Landau/Pfalz, Altes Stadthaus

HUNGARY

Joseph Durst, Szekacs J. u. 37, 5900 Oroshaza

CZECHOSLOVAKIA

Ing I. Adamovic, Suhvezdna 10, 829 00 Bratislava, CSSR

USA and CANADA

Selected Glads Inc, PO Box 26, New Albany, Ind 47150, USA

Editor, NAGC *Bulletin*, 95 Cox Ave, Armonk Village, NY 10504

International Correspondent, NAGC, George Webster, RFD, Chestnut Ridge, Glens Falls, NY 12081

Sources of supply of gladiolus corms and seeds (W) = *wholesale only;* (R) = *limited to their own introductions, seedlings, seeds etc:*

Arenius, Arthur, 123 Western Drive, Longmeadow, Mass 01106

Baldridge Gladiolus, 1729 19th Ave, Greeley, Colo 80631

Beau-Kay Gladiolus Gardens, Rt 1, Box 133, Vicksburg, Mich 49097

(R) Bebenroth Gladiolus, 10806 Ridge Rd, North Royalton, Ohio 44133

Bevington Greenacres, Galveston, Ind 46932

(R) Buch, Philip O., Rd 1, Broad Acres, Flemington, NJ 08822

(R) Buell, Clifford D., 3885 Tuscarawas Rd, Beaver, Pa 15009

Butt, Leonard W., Huttonville, Ontario, Canada

Canada Bulb & Plant Breeders, PO Box 346, Kelowna, BC, Canada

(R) Cartmell Gladiolus, 395 Berea St, Berea, Ohio 44017

Champlain View Gardens, South Hamilton, Mass 01982

(R) Colorado Hybridizers, c/o Lee Ashley, 310 S Glencoe, Denver, Colo 80222

(R) Coon, Lynn, Rt 1, Box 9, Paul, Idaho 83347

Dandurand, Paul, Rt 2, Box 252, Momence, Ill 60954

Darst Bulb Farms, PO Box 81, Mt Vernon, Wash 98273

(W) Davids & Royston, 5256 W Washington Blvd, Los Angeles, Calif 90016

Eden Gladiolus Gardens, Box 7, Mt Eden, Calif 94557

(R) Feltons Nurseries, 4349 Gladwyn Ave, Pennsauken, NJ 08110

Ferncliff Gardens, Hatzic, BC, Canada

(R) Fisher, Ted L., 5402 Village Way Court, Austin, Tex 78745

Flads Glads, 2109 Cliff Court, Madison, Wis 53713

Gondek's Glad Gardens, Daggert, Mich 49821
Gruber Glad Gardens, 2910 West Locust St, Davenport, Iowa 52804
Heart's-Ease Farms, RD Califon, NJ 07830
(W) House of Spic & Span, PO Box 63, Newfield, NJ 08344
(R) Idaho Ruffled Gladiolus Gardens, 612 E Main St, Jerome, Idaho 83338
Jackson & Perkins, Medford, Ore 97501
J.J.K. Flower Bulb Inc, PO Box 734, Upper Montclair, NJ 08344
(R) Labrum, Miles C, 761 East 6400 South, Salt Lake City, Utah 84107
Melk & Son, George, Plainfield, Wis 54966
Morans Glad Gardens, Rt 3, Box 480, Chechalis, Wash 98532
Mountain View Glads, Rt 2, Balton Lane, Boise, Idaho 83702
Nichols, Bernard E., 4827 SE 51st, Portland, Ore 97206
Noweta Gardens (Carl Fischer), St Charles, Minn 55972
Pleasant Valley Glads, 163 Senator Ave, Agawam, Mass 01001
Plummer Glad Gardens, Rt 1, Carrol, Ohio 43112
(W) Quality Gladiolus Gardens Inc, PO Box 458, Jonesboro, Ark 72401
Rich Gladiolus, Marion, NY 14505
Rupert, Laurence, Sardinia, NY 14134
Roberts, Winston, Box 3123, Boise, Idaho 83703
(W) Schipper & Co, PO Box 141, Harrison, NY 10528
Squires Bulb Farms, 3419 Eccles Ave, Ogden, Utah 84403
Summerville Gladiolus, RD 1, Glassboro, NJ 08028
Ed-Lor Glads, 234 South St, So Elgin, Ill 60177
Timberland Gardens, PO Box 1, Waterloo, Ore 97395
(W) Vandenburg, Peter J., PO Box 514 Ruskin, Fla 33570
Wagner, Michael, 215 S Washington, Memphis, Mo 63555
(W) Warner Gladiolus, PO Box 695, Medford, Ore 97501
(R) Zeller, Mrs John, Java, SD 57452

APPENDIX B

Glossary of Technical Terms

ABORT develop unnaturally, change form.

AERATION permit air to circulate within the soil.

ALLELE part of the genetic inheritance responsible for a specific characteristic.

ANTHER part of plant carrying male pollen.

ANTHOCYANIN chemical determining colour (lavender-red range).

ANTHOXANTHIN chemical determining colour (cream-yellow range).

APICAL nearest the apex, most central.

ASTER YELLOWS virus disease, mainly in north-west USA.

AXIS straight line or (here) single plane.

BACK-CROSSING cross from a seedling to one of its parents.

BASAL PLATE central disc on underside of corm.

BEAN YELLOW MOSAIC non-lethal virus transmitted from beans.

BIGENERIC involving two different genera (botanical sub-divisions).

Botrytis gladiolorum fungus disease. May be treated with 3 per cent tetrachloronitrobenzene (TCNB) dust immediately after lifting, dry at 29–32° C, store cool.

BRACT small leaf or scale below calyx.

BRACTEOLE similar to bract, but above calyx, between it and stem.

BULB mass of swollen leaf-bases for planting.

CALYX outer case of bud.

CAPILLARY thin hairlike 'vein' that draws up moisture.

CAROTENE pigment imparting orange colour.

CARPEL female reproductive organs.

CHLOROPHYLL complex chemical that imparts green colour but is vital for using the sun's energy to create chemical changes within the plant.

CHLOROSIS yellowing through absence of chlorophyll or interference with its efficiency.

CHROMOSOME inheritance-determining chain of genes that may be stained and detected under a microscope.

CLONE all the vegetative material deriving from a given seed.

CONTRACTILE ROOTS specialised roots for drawing the corm down to its optimum depth as it develops.

COROLLA the inner whorl of three petals.

CORM swollen stem-base for planting.

CORMLET spawn in the form of a tiny corm growing as one of a cluster attached to the new corm.

CUCUMBER MOSAIC virus, mainly in North America but occasionally found in Europe, causing streaks to foliage and flowers.

CULTIVAR a compound word from 'cultivated variety' used to distinguish the clone from natural varieties found in the wild. This is the botanist's way of signifying that man has had a hand in the breeding.

Curvularia trifolia a fungus that causes 'damping off', mainly among cormlet plantings.

DE-EYEING removing sprouts from corms.

DELPHINIDIN chemical responsible for blue colouring in flowers.

DIPLOID having two identical sets of chromosomes.

DORMANCY period of rest between growth cycles.

DRESSING final manipulation to flower-spike to improve its appearance.

ECOTYPE variant of a species peculiar to a particular area.

EMASCULATION removal of the male reproductive organs.

EMBRYONIC in its earliest stage of growth.

EPIDERMAL on the outer layer of skin or tissue.

FALLOW left without a planted crop during a growing season.

FERTILISATION the union of the male gamete with the female gamete, to produce a viable seed.

FILAMENT slender fibre supporting the anther.

FILIFORM ROOTS feeding roots that branch out into tiny hair-like ends.

FLOCCULATE adhere to form larger 'crumbs' of soil.

FRIABLE crumbling easily, easily worked.

FUNGUS mould, usually disease-bearing.

Fusarium oxysporum infection that enters through the roots and causes premature yellowing.

GAMETE a split cell containing only half the normal number of chromosomes.

GENE a basic unit of inheritance, forming part of a chromosome.

GENUS a particular member of a plant family, as *Gladiolus* is a genus of the Iridaceae.

GLAUCOUS covered with a surface bloom.

HANDLE vernacular term for the bare stem below the blooms on a cut spike.

HARD PAN compacted layer impervious to water.

HUMUS soil condition conducive to the increase of beneficial bacteria and the earthworm population.

HYBRID result of a cross between two plants of different clones.

INHIBITOR something that prevents or restrains normal processes.

INTERNODES lengths of bare stem between leaves.

LACINIATED cut into at the edges, giving a fringed effect.

LARVAE grubs, immature insects hatching from eggs.

LATERAL sideways, to the side.

LESIONS damaged areas, morbidly changed texture.

LOAM well-balanced soil, between sand and clay, including decayed vegetable matter.

MALVIDIN chemical producing violet and purple colouring in flowers.

MEIOSIS splitting of cell into halves with only half the normal number of chromosomes to form a gamete.

METABOLISM cycle in which nutritive material is turned into living matter and protoplasm is broken down into simpler substances.

MILDEW minute destructive fungi caused by dampness.

MITOSIS cell-division in which both halves have full complement of chromosomes; the process of cell multiplication.

MULCHING top-dressing the soil to conserve moisture by reducing evaporation and to smother weeds.

NECTAR sweet fluid produced in the throats of flowers.

NODES points on the stem from which the leaves sprout.

OUTCROSSING hybridising two unrelated or very distantly related plants of the same genus.

OVARY female reproductive organs where the ova originate.

PAPILLAE minute 'hairs' on the stigma, designed to capture pollen.

PERIANTH the total petals.

pH a measure of the acidity or alkalinity of the soil, neutral being pH 7.

PHLOEM ducted tissue for distributing sap.

PHOTOPERIODICITY length of exposure to light, which can control growth and blooming time.

PHOTOSYNTHESIS chemical reaction induced by light, particularly sunlight, releasing energy as it takes place.

PHYTOSANITARY concerned with plant health.

PISTIL total female organs of flower.

POLLINATE place pollen on to the stigma to effect a cross.

POLYPLOIDY multiples of chromosome-sets, number unspecified.

PRECIPITATE reject from solution.

PRIMORDIA first or embryonic formation.

PRIMULINUS used of hybrids having hooded flowers inherited ultimately from *G. primulinus*, now thought to be *G. nebulicola*.

PROTEIN albuminoid; basic food-chemical in plants.

RADICLE first rootlet.

ROOT NODULES tiny bumps around basal plate from which roots sprout.

RUDIMENTARY in earliest stages of development.

Sclerotinia fungus that attacks neck and roots of gladioli and is commonly known as 'dry rot'.

SECONDARIES lateral spikes attached between the foliage and the main flowerhead.

SEPALS outer petals forming calyx or all petals—botanists are not agreed among themselves (see Chapter 2).

Septoria fungus disease that attacks the leaves initially, commonly called 'hard rot'.

SIBBING crossing sister-seedlings of the same parentage.

SIDE-DRESSING sprinkling fertilisers etc in bands on the soil beside the plant-rows.

SPECIES a distinct botanical type found in the wild.

SPORE disease-bearing micro-organism carried by the wind.

SPORT common term for a natural mutation; a change within a clone rather than one achieved by breeding.

SPROUTING early growth of stems, especially before planting.

STAMEN complete male reproductive organ.

STIGMA lobe of female reproductive organ, bearing pollen-receptors.

STOLON feeding-tube from corm to cormlet.

STOMATA breathing pores on leaf-surfaces.

Stromatina form of 'dry rot' fungus that attacks neck of gladiolus.

STYLE supporting filament for the stigma.

Taeniothrips simplex scientific name for thrips, a minute flying insect, banded white and black, that attacks gladioli.

TETRAPLOIDS plants having four sets of chromosomes.

TILTH finely crumbled, well-worked top-soil.

TOBACCO RING SPOT relatively harmless disease that attacks gladioli mainly in the USA, creating pale discs on foliage.

TRANSPIRATION evaporation of moisture through plant surfaces.

TRIPLOID plant having three sets of chromosomes.

VIRUS disease-transmitting organism that in some stages acts like a living cell, in others appears inert.

XANTHOPHYLL chemical causing yellow pigmentation in plants.

XYLEM ducted tissue carrying water and minerals to the leaves.

APPENDIX C

Selected List for Further Reading

GLADIOLI IN GENERAL

Genders, Roy. *Gladioli and the Miniatures* (London: Blandford Press, 1961)

The Gladiolus Annual (yearbook of the British Gladiolus Society). 1926–

Koenig, Noreen and Crowley, Bill (eds). *The World of the Gladiolus* (Edgewood, Md, USA: Edgewood Press, 1972)

North American Gladiolus Council Bulletin (quarterly)

SOUTH AFRICAN SPECIES

Delpierre, G. R. and Du Plessis, N. M. *The Winter-growing Gladioli of South Africa* (Cape Town: Tafelberg, 1974)

Lewis, G. J., Obermeyer, A. A. and Barnard, T. T. *A Revision of the South African Species of Gladiolus* (Cape Town: Purnell, 1972)

SOIL CONDITIONING

Russell, Sir E. John. *The World of the Soil* (London: Collins, 1957)

Shewell-Cooper, W. E. *Compost Gardening* (Newton Abbot: David & Charles, 1972)

Acknowledgements

I have found this the most difficult part of the whole book to write. No one can learn a great deal about any subject from personal experience alone. I am, of course, heavily indebted to others for their contributions to gladiolus literature, for their letters concerning the gladiolus, for points picked up in conversation or from more formal 'talks', and particularly from their practical example of staging worthy exhibits at shows. Nevertheless, to try to list all by name would only lead to my giving unintentional offence by inadvertently omitting someone by oversight.

May I be permitted the simplest and perhaps most sensible way out of my difficulty? I wish to thank *en bloc* the numerous contributors over the years to *The Gladiolus Annual* and the *July Bulletin* of the British Gladiolus Society, to the quarterly *Bulletin* of the North American Gladiolus Council, and to *The World of the Gladiolus*, particularly the authors of its more scholarly chapters.

Some names, however, I feel obliged to mention and particularly wish to do so. Most valuable and instructive to me have been the friendship over the years of Mr Frank W. Unwin, the co-operation of Mrs Noreen Koenig, and that remarkable one-man publishing effort of Mr George Webster, 'SPLASH'. I am also grateful for time spared and hospitality extended by the Dutch firms of Konijnenburg en Mark, H. J. Salman & Zonen, and P. Visser Cz.

Several American growers have not only been welcome correspondents, but have also been generous towards me with their stock, particularly Mr Winston Roberts and Mr Robert Euer. For the same reasons, I am equally indebted to Mr A. N. Gromov of the USSR.

The number of BGS members that have contributed to my knowledge of, and pleasure in, the gladiolus is so vast that it would be invidious to single out individuals here—my grateful thanks to all.

I know that it seems trite to dedicate a book to one's wife and to employ the hackneyed phrase 'without whom it could never have been achieved'. In this case, it is not only the book, but much of the gardening and hybridising that would never have been achieved but for the patient and reliable assistance of my wife. Only my son and I know what her contribution has been to my life in the gladiolus world. So, to Mary, with all my love, I humbly dedicate this book.

Index

Omitted are oft-repeated words such as bloom, bud, cell, corm, cross, flower, foliage, garden, gladiolus, hydbrid, leaf, spike, and all names of countries. For main references the page numbers are printed in **bold** type.

Date Due
